PROJECT MANAGEMENT

PROJECT MANAGEMENT

Planning and Implementation

Bennet P. Lientz

Kathryn P. Rea

Harcourt
Professional Publishing

SAN DIEGO   NEW YORK   CHICAGO   LONDON

The publisher has not sought nor obtained approval of this publication from any other organization, profit or not-for-profit, and is solely responsible for its contents.

Printed in the United States of America.

ISBN: 0-15-607010-3

99   00   01   02   M&G   4   3   2

CONTENTS

Part II: Life Support for Your Project—Communications

Part III: Managing the Work

Part IV: Project Crisis

NEW IN THIS EDITION

You will find the following in the 2000 edition of this book:

- New chapter on collaborative scheduling and work
- New chapter on achieving success using e-commerce in project management
- New material on project costing
- Restructured chapter on dealing with crisis

The disc material, which has been arranged by chapter, contains new material, including the following:

- An issues-tracking spreadsheet
- A checklist for problems in projects
- Project templates for specific vertical markets

PREFACE

Project management has been a focus of many books and articles over the past half century. While technology and the nature of projects have changed, project management methods and tools described in many sources are rooted in the 1950s and 1960s. The business situation was different then, most often involving single projects with dedicated resources. Many people were available. Today's business world consists of multiple projects with shared resources. Technology has created opportunities for better managing projects, such as using e-commerce. This book is based on modern proven methods and tools. It is about dealing with projects of the 21st century.

The goal of the book is to help you be successful in projects. Therefore, the focus is on project issues, opportunities, and crises. Our cumulative experience of managing more than 75 projects and participating in more than 150 others has produced a wealth of lessons learned and methods that are passed on to you in this book.

The perspective of the book is from a business as well as a technical worldview. To be successful in projects you must consider management, politics, organization, and business processes, as well as technology. In the real world your project competes for money, resources, and management attention. You must compete for funding, management support, organization involvement, and resources. You have to do this while dealing with competing demands as well as change in organization, policy, management, technology, and the marketplace.

The book is organized around the steps in a project—from the initial conception of a project idea to the successful end of the project. Some chapters address specific important areas such as multiple projects, collaborative scheduling and work, dealing with crises, and projects that use electronic commerce, a major force of change affecting companies and firms in all business segments. To make the material easy to read and fast to use, chapters follow a standard structure with key strategies, directions, and examples. Examples come from a variety of industries, including banking, consumer products, manufacturing, transportation, and construction.

Who can benefit from this book? People who either are currently involved in projects or potentially are going to be—managers, professionals, and support staff. Those who have successfully used the materials in the book have ranged from individuals with little or no project experience to professionals with many years of project management experience looking to hone their skills. The material has been developed over the past seven years and has been employed in more than 80 organizations worldwide.

Some critical tools you will receive in the book and on the CD-ROM are the following:

- Full examples of projects from a variety of industries
- More than 250 lessons learned and guidelines
- A step-by-step approach for setting up a project
- Information on how to cope with more than 100 project issues
- Information on how to manage multiple projects with resources spread across projects and while having to do normal work
- Electronic templates on disk
- Checklists in both electronic and paper form

- Down-to-earth tips on using the Internet, groupware, and other software for project management
- Instruction on how to manage electronic commerce projects
- Instruction on how to manage projects in a collaborative way for greater success
- Advice on how to prevent project problems and deal with project crises

150 projects ago, we desperately wanted a book that gave real-world help. 150 projects later, we are pleased to place this book in your hands.

Bennet P. Lientz
Kathryn P. Rea

ABOUT THE AUTHORS

Bennet P. Lientz is a Professor of Information Systems at the Anderson Graduate School of Management, University of California Los Angeles. Dr. Lientz was previously Associate Professor of Engineering at the University of Southern California and department manager at System Development Corporation. He managed and reengineered administrative systems at UCLA and has managed a variety of projects since the late 1970s.

Professor Lientz has taught project management since 1975. He has delivered project management seminars to more than 2,500 people in Asia, North America, Europe, and Latin America. Dr. Lientz is the author of ten books and more than 50 articles on business process improvement, project management, and information systems.

Kathryn P. Rea is president and founder of The Consulting Edge, Inc., which was established in 1984. The firm specializes in business process improvement, project management, and information systems.

Ms. Rea has managed more than 55 major projects internationally. She has worked with industries such as government, energy, banking and finance, insurance, distribution, manufacturing, import/export, global trading, and retailing. She has successfully directed multinational projects in South East Asia, China, North America, Latin America, and Europe. Ms. Rea has conducted numerous seminars on project management and business process improvement.

ABOUT THE COMPUTER DISC

SYSTEM REQUIREMENTS

- IBM PC or compatible computer with CD-ROM drive

- WordPerfect® 6.0 for Windows™, Microsoft Word® 6.0 for Windows™, or compatible word processor, and Microsoft Excel® 5.0 for Windows™ or compatible spreadsheet, Microsoft® PowerPoint 7.0, Microsoft® Project 98, and Microsoft® Access

- 3 MB available on hard disk/drive

The CD-ROM provided with *Project Management,* 2000 edition, contains files to support use of the steps of project management presented in this book.

The word processing forms are intended to be used in conjunction with your word processing software. The word processing forms have been formatted in Word 6.0 for Windows and WordPerfect 6.0 for Windows versions. If you do not own either of these programs, your word processing software may be able to convert the documents into a usable format. Check the user's manual that accompanies your word processing software for more information about the conversion of the documents.

Also included on the disc are Microsoft PowerPoint® 7.0 presentations, Microsoft Project® 98, Microsoft® Access, and Microsoft Excel® 5.0 for Windows™ or a compatible spreadsheet.

Subject to the conditions in the license agreement and the limited warranty, which is displayed on screen when the disc is installed and is reproduced at the end of the book, you may duplicate the files on this disc, modify them as necessary, and create your own customized versions. Installing the disc contents and/or using the disc in any way indicates that you accept the terms of the license agreement.

If you experience any difficulties installing or using the files included on this disc and cannot resolve the problem using the information presented in this section, call our toll-free software support hotline at (888) 551-7127.

INSTALLING THE TEMPLATES

To install the files on the disc using Windows 95 or above, select the Control Panel from the Start menu. Then choose Add/Remove Programs and select Install. You will be asked a series of questions. Read each question carefully and answer as indicated. If you are using Windows 3.1, choose File, Run from the Windows Program Manager and type D:/INSTALL in the command line or type D:/INSTALL at the DOS prompt.

First, the installation program will ask you to specify which drive you want to install to. You will then be instructed to specify the complete path where you would like the files installed. The installation program will suggest a directory for you, but you can name the directory anything you like. If the directory does not exist, the program will create it for you.

You can choose to install the Excel, PowerPoint, Project and either word processing program (Word or WordPerfect). The program will automatically install the files in Excel, PowerPoint, Project, Word, or WordPerfect subdirectories.

OPENING THE FILES

Open your word processing program. If you are using Microsoft Word or WordPerfect 6.0, choose Open from the File menu. Select the subdirectory that contains the loaded files to list the names of the files. Highlight the name of the file you want to open and click OK or press ENTER. You can also open a document in Windows Explorer (in Windows 95 or above), or in Windows 3.1 from the File Manager, by highlighting the name of the file you want to use and double-clicking your left mouse button.

Refer to the Disc Contents section of the book to find the file name of the document you want to use. The Disc Contents are also available on your disc in a file called "CONTENTS." You can open this file and view it on your screen or print a hard copy to use for reference.

Opening the Excel Spreadsheet

Open Excel 5.0 or higher or a compatible spreadsheet program. Choose Open from the file menu. Select the subdirectory that contains the loaded files to list the names of the files. Highlight the name of the file you want to open and click OK or press ENTER. You can also open a spreadsheet file from the File Manager (in Windows 3.1) or in the Explorer (in Windows 95) by highlighting the name of the file you want to use and double-clicking your left mouse button.

Spreadsheet Tips

Initially, only one spreadsheet file will exist in the spreadsheet installation directory. This is done to preserve an original copy of the Excel spreadsheet. After opening this file, simply select File and save to a new location and/or give the spreadsheet a different name. This will create a new file that will not be marked as read-only.

Opening the PowerPoint Tips

Open PowerPoint 7.0 or higher or a compatible program. Choose "Open an existing presentation" from the file menu. Click OK or press ENTER. Select the subdirectory that contains the loaded files to list the names of the files. Highlight the name of the file you want to open and click OK or press ENTER. You can also open a PowerPoint file from the File Manager (in Windows 3.1) or in the Explorer (in Windows 95) by highlighting the name of the file you want to use and double-clicking your left mouse button.

Opening the Project Tips

Open Project 98 or higher or a compatible program. Choose Open from the file menu. Select the subdirectory that contains the loaded files to list the names of the files. Highlight the name of the file you want to open and click OK or press ENTER. You can also open a Project file from the File Manager (in Windows 3.1) or in the Explorer (in Windows 95) by highlighting the name of the file you want to use and double-clicking your left mouse button.

Word Processing Tips

Wherever possible, the text of the documents has been formatted as tables so that you can modify the text without altering the format of the documents. To maneuver within a table, press TAB to move to the next cell, and SHIFT + TAB to move backward one cell. If you want to move to a tab stop within a cell, press CTRL + TAB. For additional tips on working within tables, consult your word processor's manual. It might be helpful to turn on the invisible table lines in Microsoft Word while modifying the document by selecting Gridlines from the Table menu. In WordPerfect, select Reveal Codes from the View menu to reveal all formatting codes at the bottom of the screen; this will help you to determine the shape of the table.

Microsoft Word and WordPerfect are equipped with search capabilities to help you locate specific words or phrases within a document. The Find option listed under the Edit menu performs a search in both Microsoft Word and in WordPerfect 6.0.

Important: When you are finished using a file you will be asked to save it. If you have modified the file, you may want to save the modified file under a different name rather than the name of the original file. (Your word processing program will prompt you for a file name.) This will enable you to reuse the original file without your modifications. If you want to replace the original file with your modified file, save but do not change the name of the file.

PRINT TROUBLESHOOTING

If you are having difficulty printing your document, the following suggestions may correct the problem:

Microsoft Word

- Select Print from the Microsoft Word File menu. Then choose the Printer function.
- Ensure that the correct printer is selected.
- From this window, choose Options.
- In the media box, make sure that the paper size is correct and that the proper paper tray is selected.

- Check your network connections if applicable.

- If you still have trouble printing successfully, it may be because your printer does not recognize the font Times New Roman. At this point, you should change the font of the document to your default font by selecting the document (CTRL + A) and then choosing Font from the Format menu and highlighting the name of the font you normally use. Changing the font of the document may require additional adjustments to the document format, such as margins, tab stops, and table cell height and width. Select Page Layout from the View menu to view the appearance of the pages before you try to print again.

WordPerfect

- Select Print from the WordPerfect 6.0 File menu. Then choose Select.

- Make sure the correct printer is selected.

- From this menu, press Setup.

- Ensure the correct paper size and paper source are selected.

- You may be having difficulty because your printer does not recognize the selected font. You can correct this problem by changing the base font of the document to your default font. From the Edit menu choose Select All (or press CTRL+A). The entire text of the document should be highlighted. Then choose Font from the Layout menu and highlight the font you normally use. Changing the font of the document may require additional adjustments to the document format, such as margins, tab stops, and table cell height and width. Select Two Page from the View menu to view the appearance of the pages before you try to print again.

DISC CONTENTS

Document Title	Filename	File Type

Chapter 1

Current Projects	01-01	Word processing
Dimensions of Project Management	01-02	Word processing
Potential Problems with Projects	01-03	Word processing
Project Evaluation	01-04	Word processing

Chapter 2

Development of the Plan	02-01	Word processing
Project Concept and Initial Planning	02-02	Word processing
Project Concept Presentation	02-03	PowerPoint
Custom Project Template	02-04	Project
Project Template for New Product Development	02-05	Project
Database—Issues	02-06	Access
Database—Lessons Learned	02-07	Access
Issues Database	02-08	Excel
Lessons Learned Database	02-09	Excel
Software Project Template	02-10	Project

Chapter 3

Project Leader Evaluation	03-01	Word processing
Project Manager Evaluation	03-02	Word processing
Support for Project Managers	03-03	Word processing
Project Management Assessment	03-04	Word processing

Chapter 4

Team Member Project Evaluation	04-01	Word processing

Chapter 5

The Project Budget	05-01	Word processing
Sample Common Resource Pool	05-02	Project

PART I ONE

GETTING STARTED

CHAPTER 1

PROJECT MANAGEMENT IN A NUTSHELL

CONTENTS

1

PROJECT MANAGEMENT IN A NUTSHELL

INTRODUCTION

Over the years, successfully managed projects have come to require a distinct set of procedures, rules, and policies. The phrase *project management* refers to these procedures, rules, and policies, as well as to style of management, politics, and other factors that go with managing projects.

Fifty years ago project management was not a formal area. People carried out projects without formal tools and methods. With the increasing use of project management in the military and in construction, methods and tools emerged, along with techniques. These became quite formalized. Computers provided software tools to support these methods. Until the early 1990s, projects were locked into this overly-formal approach.

As project management grew to be a recognized area, problems arose. For example, it was difficult to manage small and large projects in the same way. Also, some managers were discouraged from participating in project management due to the overhead. People wanted just to plunge in and do the work. This approach will often succeed for small projects which encounter no problems. However, if you run into an issue or problem, then you have no structure for control and management.

The modern approach to project management emerged in the 1990s, a transformation due to changing business needs and to the availability of new electronic tools. The widespread use of computer networks has had a significant impact on the practical aspects of project management, along with the Internet, intranets, electronic mail, shared project management software, databases, and groupware. In the twenty-first century, more projects will be handled in a collaborative way, since people who are working on the projects will be able to access the project data easily to update and comment on the tasks, schedules, issues, and lessons learned. Using networking and the Internet, people can share information and make decisions related to a project across thousands of miles. This electronic approach to project management will be a key theme of this book.

Let's contrast the old methods with the modern approach in a table. The business side of projects has changed. There are more projects and fewer people qualified to do the work. This problem is magnified by increased management pressure due to the importance of the project. The following table shows the contrast between the old methods and the modern approach. In this table you see the word *scheduler*. This refers to someone who coordinates projects and supports a project team. It may or may not be the project leader. In larger organizations with formal project management, the scheduler typically belongs to an organization called *the project office*, which is the home of the schedulers who go out and coordinate various projects.

Characteristic	Earlier Project Management	Modern Approach
Focus	Schedule updating and status checking	Issue management and action item orientation
Software tools	Project management	Project management, groupware, electronic mail, Internet, databases
Project management software	Single user	Multiple users
Role of the scheduler	Project administrator	Training and analysis support

Why do problems arise in managing projects? One answer is that a project is different from standard work. A project is more focused. It generally gets more management attention because of this and because the goal is of direct importance to management. Risk is involved, and there is little opportunity to hide failure. Sometimes organizations attempt to carry out projects without calling them projects. Structure is lacking. If you are going to use project management, take it seriously and follow project management methods.

In general, project managers have not made an effort to apply lessons learned during the past decades of project management. Lessons learned are one of the cheapest and easiest improvements to make. This book will emphasize strategies learned as a result of experience. Directions are given on how to deal with commonly encountered situations. Without relying on lessons learned, people repeat mistakes needlessly.

PURPOSE AND SCOPE

This book offers guidance on how to perform and manage projects quickly, effectively, and successfully. Experience shows that projects of all types over time and across many industries share much in common. The expertise of project managers worldwide is garnered in this book to enable you to get a quick start and achieve success. The fundamentals of project management in the political world will also be covered—how to solve problems, take advantage of opportunities, and manage resources.

At the end of each chapter, key strategies learned during the development of project management to its present form will be given. These are based on the project experiences of more than 100 companies over the past 25 years; these are proven key strategies to help you be a successful project manager.

Chapter 19 will focus on 100 commonly encountered problem situations. Examples from industry are given. For each problem, you can learn how to prevent failures.

The scope of the book spans the entire life cycle of a project—from concept to follow-up after a project is finished. Projects of all sizes are considered in different industries and areas of the world. Methods and tools discussed range from manual to automated.

Potential and current project leaders and managers will learn how to address key issues and opportunities in a project and how to be more efficient and effective as leaders. Project team members will learn how to contribute more effectively to the team and project and how to participate in project management. Managers will learn how to set up, oversee, and direct multiple, diverse projects more effectively. They will understand the critical ingredients for success and how to avoid project problems.

LINE ORGANIZATION AND PROJECT MANAGEMENT DIFFERENCES

A project is the allocation of resources to achieve a specific set of objectives following a planned and organized approach. A project has a focus—specific end results. Organizations form project teams to address the project work, while routine work is performed by the line organizations. Putting a new sprinkler system in the back yard is project work. Turning on the sprinklers and maintaining the lawn and garden is routine work.

Line organizations have specific rules and procedures. Some apply to projects as well; others do not. A project represents a separate structure outside of the line organization or functional departments. This puts the spotlight on it and creates tension between the line organization and the project organization. When you employ people from a series of line organizations in various projects, you have a *matrix organization*. Aerospace and other firms have employed such organizations for years. Today, an individual can be assigned to several projects as well as line organization work. For example, a network support person may be implementing a new network in one location, upgrading the network in a second location, and handling troubleshooting—all at the same time. A marketing product manager could be the project leader for a new product roll-out as well as reporting on existing product sales.

When line organization techniques are applied to projects, the projects fail. Line organization techniques do not provide sufficient focus and support for the project. Attention goes mainly to routine work and the best people are often kept doing standard work.

Here are some key differences between line organizations and project management:

Comparison	Line Organization	Project
Work characteristics	Routine, stable	Driven by milestones
Duration	Almost eternal	Temporary
People	Well defined roles	Roles depend on project tasks and can change
Projects favor	Entrenched staff	Younger staff
Potential for recognition	Limited	Can be substantial
People and risk	Often risk averse	Often risk takers

ESSENTIAL PROJECT MANAGEMENT CONCEPTS

Before plunging into the project plan and project management, let's review some project-related concepts to provide a common basis for understanding.

The Schedule

A plan or *schedule* for a project consists of the following:

- Tasks (detailed activities)—each has a start date, an end date, and a duration
- Summary tasks—a rollup of several detailed tasks (i.e., an outline format)
- Milestones—deliverable items and end products of the project; a milestone is usually the last item under a summary task
- Dependencies between tasks—two tasks can depend on each other in a number of ways: head-to-tail (one task cannot start until another is completed), lag (one task must start five days after another task), or lead (one task must start 10 days before another task)
- Resources—personnel, facilities, equipment, and other resources that are assigned to tasks
- Calendars—the project, tasks, and resources can all have separate calendars

Let's suppose that you want to plant a garden. This is a *summary task* at the highest level. Digging up the ground is a detailed task under the summary. The *milestone* is that the ground is ready for fertilizing and watering. Both of these tasks depend on the ground being prepared first. If you do the work yourself, your *resources* are yourself, a shovel, and a pick axe. You will have to borrow a special cultivator tool and you can only use it a few hours a day. It must be returned each day. This resource has a *calendar* different from yours and your equipment's.

In some schedules, tasks may have four or five sets of dates. Baseline plan dates (defined below), customer dates, project manager dates, and actual dates are four commonly used dates. Compare schedules based on different dates.

The Task

For each *task*, generate the following:

- Task duration—typical elapsed number of working days
- Task calendar—allow a task to be worked different hours and days
- Resource and project calendar—working hours and days
- Start and finish dates—tasks may have many different sets of dates
- Task constraint—how the task is scheduled (as soon as possible, must start on a specific date, as late as possible, must end on a specific date, etc.)
- Resources assigned to the task

Critical Path

If you trace a series of lines from the start of a project to the end, this is called a path. The ***critical path*** or ***mathematical critical path*** is the longest path in the project. It is important because if anything is delayed on the critical path, the project is delayed. Any task can happen to be upon the mathematical critical path (even holding a meeting) if the durations, dependencies, and dates place it there.

Slack time refers to the amount of time that a task can slip before it becomes part of the critical path. The time a task can slip before it impacts another task is called *free slack*. The time that it can slip before the project slips is called *total slack*.

RISK AND ISSUES

You have heard people say, "This project has risk," or "This task has risk." They mean that there is a substantial likelihood that the project or task will run into trouble. The schedule overall may slip. The project may fail. To say that a task or project has risk is to indicate that underlying issues are associated with the task or project that must be resolved to prevent trouble.

You can associate tasks with risk with a list of issues. A task can have multiple issues; an issue can apply to several tasks. It is important to note that *managing risk means managing issues*. If you don't resolve an issue, the tasks start to slip and the project falls behind. There can be political impact as well. The problem moves from managing problems and issues to managing crisis. This escalation from problem to crisis is a major focus of this book.

Management Critical Path

With many project problems has come a realization of the importance of considering risk. The *management critical path* is any path in the project that involves tasks with high levels of risk. If you manage these paths, you manage risk. Tasks on this path should take priority. You can go back to giving priority to managing the mathematical critical path later. As a result of keeping the management critical path in control, you reduce your overall project risk.

Multiple Projects

How do you deal with *multiple projects*? There is both a tactical and a strategic approach. On a tactical level, if you define a number of different projects and save them in individual computer files, you can then combine them into a single schedule (merge the schedules). Or, you can cut and paste from each into a general schedule. A third approach is to have a general high-level schedule reference the detailed schedules so that an individual task in the general plan is a rollup of the entire detailed schedule (rollup). These are three ways to combine a number of projects.

Moving to strategy, the problem is how to manage the resources that are shared among projects and non-project work. Where do you place the priority? How do you manage issues that apply to multiple projects? How do you address dependencies among different projects? How do you get an overall view of the projects? These are all critical questions today. If you don't have some minimal level of standardization among the projects, doing multiple project analysis is very difficult.

Project Template

A *project template* refers to the task list at a high level, resources, dependencies, and assignment of resources to tasks. The template does not include durations, dates, or the number of each resource for each task. A *work breakdown structure* (WBS) includes just the list of tasks. The template gives more useful results than a WBS because it has more of the preparatory work completed for the project. You typically create high-level, outline tasks for the template. The lowest level of a task in a template is about one to two months. Then the individual project leader fills in the details for the specific project. This brings detailed tasks down to one to two weeks. Follow this approach to ensure a greater degree of standardization.

All projects should be based on templates so that you can manage multiple projects. A template approach has many benefits, including the following:

- Time is saved in setting up a schedule by building on the template.

- A common base of knowledge is built through the templates.

- Standardization is provided at the higher level for analysis of multiple projects and at the lower level for detail.

- Lessons learned from previous projects are used to improve the templates.

GANTT and PERT Charts

A project uses two basic types of charts: GANTT and PERT. The GANTT chart is the most useful. Figure 1.1 is a sample GANTT chart which highlights some of the complexity that is possible. At the top of the chart, the first set of tasks is the template; the lower set, beginning on line 10, are the tasks of the project. In a real project these would be in two separate projects. Here, they are on the same chart so you can see how a template works (more detail is filled in for the schedule).

Suggestions on how to use a GANTT chart:

- Place tasks in outline form; detailed tasks are the most indented.

- Use symbols to stand for tasks and milestones in the chart.

- Show three sets of dates—customer plan dates, internal business plan dates, and actual dates—to compare real and ideal.

- Number tasks so that it is easy to talk about them in a meeting without reading the task name. This allows you to add tasks within the same framework.

PERT (Program Evaluation and Review Technique) charts are best at showing dependencies (see Figure 1.2). Use this chart when defining the project plan initially to obtain agreement on dependencies. Then use the more detailed GANTT chart.

Building on a template you can construct a unique project plan for the specific project, define more detailed subtasks, allocate resources to the subtasks, define durations, and set schedules. These would then "rollup" to the summary tasks of the template.

You can now assign costs to resources, add any overhead and other costs, and compute the total estimated cost of the project. Most software programs can easily do this for you. Caution—these numbers are only estimates and are almost never correct. They don't account for complex overtime, amortization, fee for use, overhead, etc., factors that software project management programs cannot easily handle.

Baseline Plan

The created schedule is reviewed and approved. The dates will be established as targets. These target dates become the baseline schedule dates or planned dates. The plan is then the *baseline plan*. The start and finish dates of the baseline plan are locked in for later analysis. The actual start and finish dates move.

Updating the Schedule

How do you update the schedule? When you actually start the project and perform work, you will mark tasks as complete. Then you will be creating another series of dates—actual dates. You thereby create another schedule—the *actual schedule*.

Actual vs. Planned Analysis

If you compare the actual and planned dates, you obtain the *actual vs. planned analysis*. You could use yet another set of dates to model

ID	Task Name	Duration	Note	Predecessors
1	Template	1d		
2	1000 Task 1	2d	Template	
3	1100 Task 1.1	1d	Template	
4	1200 Task 1.2	1d	Template	
5	1300 Task 1.3	1d	Template	3,4
6	2000 Task 2	3d	Template	2
7	2100 Task 2.1	1d	Template	
8	2200 Task 2.2	1d	Template	7
9	2300 Task 2.3	1d	Template	8
10	Detail	1d		
11	1000 Task 1	16d	Template	
12	1100 Task 1.1	10d	Template	
13	1110 Task 1.1.1	5d	Detail	
14	1120 Task 1.1.2	10d	Detail	
15	1200 Task 1.2	5d	Template	
16	1210 Task 1.2.1	5d	Detail	
17	1300 Task 1.3	6d	Template, Detail	16,12
18	2000 Task 2	29d	Template	11
19	2100 Task 2.1	12d	Template	
20	2110 Task 2.1.1	7d	Detail	
21	2120 Task 2.2.2	5d	Detail	20
22	2200 Task 2.2	9d	Template	21
23	2210 Task 2.2.1	5d	Detail	
24	2220 Task 2.2.2	4d	Detail	23
25	2300 Task 2.3	8d	Template, Detail	24

Figure 1.1
GANTT Chart

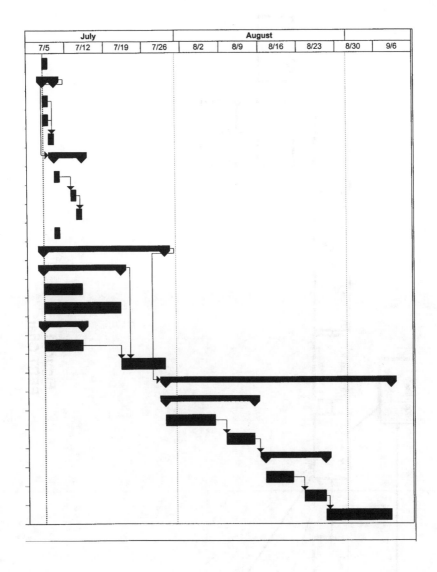

Figure 1.1, *continued*
GANTT Chart

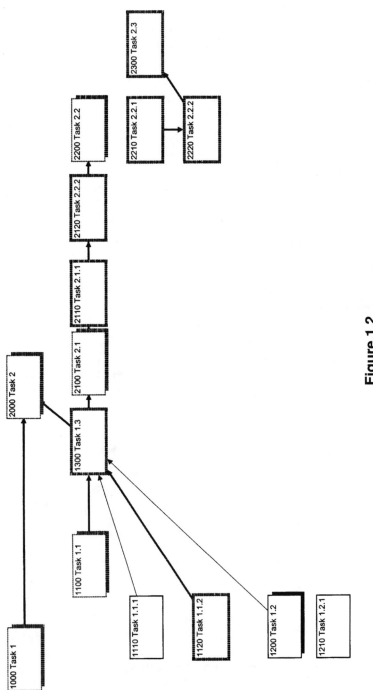

Figure 1.2
PERT Chart

"What would happen if ...?" This "What if...?" analysis is one of the benefits of project management software systems. You could get yet another set of dates if you add the dates that you promised the project to a customer.

The Project Plan

To be successful, you need more than a GANTT chart. Your *project plan* must include the following:

- Project objectives—clearly and unambiguously state the purpose of the project
- Project strategies—identify your approach for directing resources and handling issues
- A detailed project plan—the set of tasks, resource assignment, dependencies, and schedules
- Project resources—all personnel, facilities, equipment, and any other resources
- Project methods—techniques to be used in creating, maintaining, and managing the project
- Project tools—software or other products that support the methods
- A definition of what work will constitute the project
- Policies and procedures for developing, updating, and reviewing projects
- An approach for defining and resolving issues and opportunities
- Standard budgeting and cost methods
- Methods and tools to support the projects and project management
- A list of issues associated with the project and which tasks correspond to these issues (tasks that have risk)

ALTERNATIVES TO PROJECT MANAGEMENT

Not everything should be treated as a project. Here are some pluses and minuses to consider when deciding whether to initiate a project.

- **Project management pluses:**

    —Provides structure for work through established methods and tools and supported by guidelines

    —Supports the gathering of experience and lessons learned

    —Provides the necessary structure to manage multiple different or related projects

- **Project management minuses:**

    —Creates tension in the organization between line and project managers

    —Takes key people from the line organization

    —Consumes resources through overhead in meetings, tracking, etc.

Consider all of the following alternatives at the start of the project. One may be preferable to project management.

- **Alternative 1: Use the line organization.** If you start a project in a line organization, you will have to consider carefully when it should be moved outside of the organization. This approach may work for very limited projects with a scope that falls within the organization.

- **Alternative 2: Outsource the project.** Hire an outside organization to do the project. Government agencies and companies often do this for construction and other projects, because the agency or company is only interested in the end result. Choose this alternative if you can't answer the question: "Why do you want to do the project yourself?"

- **Alternative 3: Do it as a joint project with outsiders.** One benefit of this method is having an outsider participate and/or review the project and offer suggestions on issues and the project in general. Outsiders might have specific experience, equipment, or expertise to contribute to the project. Also, they can share the risk.

- **Alternative 4: Create a project organization.** Decide what pieces of work will be projects. Establish an approach for managing multiple projects. Create project templates.

Projects and Business Processes

Projects and processes are intimately linked. The goal of projects is to improve business processes. Projects consume resources and time and so they have to deliver this benefit. A business depends for its revenue and costs on its core business processes. As projects change the business processes, revenue should increase and costs should decrease.

Many projects employ technology and systems to improve business processes. Thus project management is linked to both business process improvement and information technology. A critical success factor in using technology is the selection of the technology. Strategic technology planning and management are functions of project management.

PROJECT MANAGEMENT MYTHS

Over the past 40 years a number of myths have arisen around project management. If you believe in and adhere to these myths, your projects are likely to fail.

- **Myth No. 1: Projects are unique.** Even if projects are wildly different in size, nature of work, and scope, they have many attributes of management in common. Buying into this myth makes it difficult to manage multiple projects. If you follow this myth, you will find it difficult to use lessons learned; you will be unable to use the template approach, which assumes similarity among different projects at a high level.

- **Myth No. 2: All projects are the same.** As with the other extreme, this outlook will lead to failure, too. If you manage all projects in the same way, then you will over-manage the small projects and under-manage the large projects. You want to manage based on issues and risk because that's where the exposure is.

- **Myth No. 3: Using a core of project leaders and members across multiple projects is beneficial.** While this has some appeal, this approach has a history of burning people out. They get stale. While they bring their experience, they tend to fall into habits with respect to methods and tools. Every project they get receives the same approach. Instead, transition new people into projects while retaining people with experience. Team up new members with experienced mentors. Rotate people so that lessons learned can spread through the organization.

- **Myth No. 4: Manage based on size and scope.** This is widely used and does have some validity, especially in construction and other industries. However, this approach leaves out the most important factor in a project—risk. When you manage projects, you want them to be successful. By focusing on size alone, risk can be overlooked. A big project does not necessarily equate to high risk.

- **Myth No. 5: You can accomplish a "death march" project through the use of tools.** Under this myth, a large project under severe constraints can succeed through the right choice of methods and tools. Though tools can help the project, they can also slow it down. The "death march" (a phrase used in World War II to refer to a person wearing out) should never have started in the first place.

SUCCESSFUL PROJECTS

Successful projects have these factors in common:

- **Success Factor No. 1: The project has a regular flow of interim, measurable milestones.** This allows management to ensure that the project overall is on schedule. People on the project team can be given a pat on the back as they made progress. Example Construction of modern military aircraft used to be done on a one-shot basis; now it is funded in stages.

- **Success Factor No. 2: The project team can tolerate change and issues.** The project team is adaptive to internal and external change. In such projects the staff has a positive attitude toward change.

- **Success Factor No. 3: Management avoids micromanagement and blanket approval.** Management monitors progress and evaluates the milestones or end products of the project.

- **Success Factor No. 4: The purpose and scope are clearly defined.** Clarity of definition varies with the type of project. Example Manufacturing, distribution, and construction projects tend to be well defined. Software, new product, and general analysis projects are more prone to problems. In part, this is due to a lack of clarity in purpose and scope. The American military learned a hard lesson about clarity of purpose from Vietnam and applied it to the Gulf War and other engagements.

- **Success Factor No. 5: The team works together in a collaborative way.** Information is shared; there is no hiding of information. The team members participate in some project management tasks, such as defining, estimating, and updating their own tasks.

UNSUCCESSFUL PROJECTS

Common reasons for failure include the following:

- **The original purpose changed, but the project didn't.** The resulting work goes unused. An example is a continuing project for a weapons systems or nuclear power plant for which the original need disappeared.

- **The project was mismanaged.** Milestone dates were missed; morale dropped; people still produced PERT and GANTT charts. Issues and resources were not properly managed. Mismanagement can occur in any area of project management and can effect the entire project.

- **The project scope changed often.** People get confused by changing requirements and sense chaos. Progress grinds to a halt.

- **Project goals were not realistic.** For example, one project attempted to model world political conditions through computer simulation. This went on for three years and cost millions of dollars. The project was finally terminated.

- **The elapsed time for the project was too long.** There was a lack of intermediate milestones. The project churned up resources and never delivered. This occurred in a large insurance project that cost more than $6,000,000 over a three-year period.

- **Management assumed the wrong roles.** Management either did not support the project or, at the other extreme, micromanaged. An example would be a project in which management left the project alone and then, when it fell into trouble, micromanaged it—two errors affecting the same project.

- **Management has no way to terminate a project.** Projects that should have been abandoned just seem to live too long. No policy is available to reassess or to redirect the project in an organized manner.

THE ORGANIZATION OF THIS BOOK

This book is organized to reflect the new project management environment. Part I addresses the development of the plan, Part II deals with management and communications, and Part III concerns the work on the project itself. It also includes chapters on collaborative project management, multiple projects, and electronic commerce projects. A fourth section addresses project crises and 100 things that can go wrong.

The chapters address the key managerial, technological, and organizational areas of project management. Each chapter also addresses issue management. (A separate chapter covers this topic in detail.) Special attention is paid to collaborative scheduling and the use of modern tools to help in project management, with a chapter on using the Internet and the World Wide Web for project management.

Each chapter is organized in the same basic manner. The introduction and the purpose and scope of the chapter are followed by the approach for the specific topic. This material is organized around steps and actions you can take. Later in the chapter, key strategies are given. These are specific ideas and recommendations for reducing risk and ensuring that your project has a competitive advantage over other projects vying for funding and management attention. The key strategies are followed by several industry examples.

Each chapter ends with having you take stock of where you are (Status Check) and presents specific action items for you to pursue.

The appendix provides sources of information for your later use. In addition, there is a glossary of terms.

If you are using this book in a project, you might want to select the chapter that fits where you are in the project. You also might start with the index, which is organized by topic or issue.

For a class or seminar in project management, you can proceed through the book sequentially. The flow of the chapters follows that of a project. By taking a case or a sample project you can apply the guidelines in the chapters to a realistic situation. This has been done with a combination of group and individual projects.

EXAMPLES

The Railroad Example

Railroads began in the early 1800s in England. The technology spread to the United States and by 1840, the United States had more than 50 percent of the world's railroad miles. Railroad construction was one of the largest public building projects undertaken by mankind. In the 1800s, the railroad industry and its related industries employed more people than any other industry. The railroad changed society's idea of space and time. Each railroad and its extensions were viewed as a project. Companies that were successful then exported their skills, knowledge, and lessons learned to build railways in foreign countries.

The railroad industry from the 1800s until the second decade of the twentieth century represents a pinnacle of project management. It had all of the ingredients that make projects interesting: risk, politics, intrigue, technology, and personalities. Without computers and formal tools, project plans were developed and implemented. The standards held by the railroads impacted the quality of output and nature of production of most major industries. Countries were changed forever by the projects. Railroads put the producers of goods directly in touch with buyers.

Railroad projects were very large, as shown by the following statistics:

- At one time in nineteenth century England, one firm had more than 700 miles of track under construction.

- Building bridges once took months or years, but the railroad industry was able to construct bridges in a matter of weeks. In England, more than 25,000 bridges were built in the first 15 years of railroad construction. This was more than the total number of bridges in the country prior to the railroad.

- A single railroad often had more employees than the entire federal government.

- Railroad projects consumed almost half of the investment capital in the United States prior to the Civil War.

- In most countries, railroad construction consumed 10 to 15 percent of investment capital for decades.

- A wide variation existed between countries and companies in their approaches to railroad financing and construction.

Because they were distributed, railroad projects had to involved substantial collaborative effort.

Some examples demonstrate the impact of different objectives and of the scope of a project, as well as the impact of the project on the organization and society. Through the following chapters, you will see how railroad projects were formulated, how they were managed, why they failed, and why they succeeded. Both good and poor managers (engineers) will be examined.

The railroad examples given in the following chapters have aspects that apply to almost all parts of project management. Railroads are a prime example of how people actively acquired and used lessons learned as they moved from one project to the next. A company's success was repeatedly tied to its ability to capitalize on experience and lessons learned. The railroad examples also show that many of the guidelines and rules for project management have been around for a long time.

Modern Examples

In addition to the railroad examples, several real companies will be considered throughout the book. At the end of each chapter, examples will be discussed in terms of lessons learned.

The three companies used as examples are the following:

- **A multinational corporation that has a number of facilities in Southeast Asia** It is implementing a large international network in ten countries to support manufacturing, distribution, and sales. This project has tight time deadlines. The scope of the project includes hardware and communications components, as well as application software and software tools to support communications. The multinational corporation provides excellent large-scale project examples.

- **An engineering firm that has multiple engineering and construction projects around the country and overseas** The major problems in this company do not concern the individual project, but rather concern managing multiple projects. This company has never had a standard work breakdown structure or a template schedule.

  The engineering company went from single-person project management methods and tools to network-based tools and increased productivity by more than 40 percent. Project costs dropped 20 percent through better management of resources. Work quality is higher. Schedules are shorter. Active effort is made to gather and apply lessons learned. The firm is much more responsive to the demands of the marketplace. Turnover of staff was dramatically reduced.

- **A consumer products firm that rolls out 5 to 10 new products and variations each year** In the past, the new product rollouts have been uneven. The work here is to develop a project template that can be used within families of products. However, the project leader must also coordinate with many departments to complete the deployment of the new product of the family as well as managing changes to the current product line.

These three company examples provide a mix of experience, industries, and scope of projects. The problems these companies have encountered are fairly typical of what you face in project implementation today.

GUIDELINES

Lessons learned (key strategies) appear in each chapter. Each is stated in a succinct manner and then discussed in more detail. Each is intended to reduce project risk, deal with issues, and ensure that you have a competitive project.

- **Clearly delineate the ground rules regarding when work becomes a project.**

  If you begin work and get halfway through, you have set a pattern. If you then decide to label the work a project, you would apply project management, methods, and tools to the work—retroactively. Progress will slow down while this is being established. It will also mean that the work will be redirected into a more formal approach. These are two reasons for determining that a project is a project early in development. If you wait, your risk increases, along with the likelihood of failure..

- **As a project leader, maintain a low profile with your project.**

  Once you label something as a project, the rules of project management begin to apply. Things get more formal and organized. Monies allocated to the work are more visible. This brings attention to the work. However, treating situations as projects raises visibility, perhaps raising risk. First, if the process is too formal, progress will be slowed. Second, the added visibility and attention may not be in the best interest of the organization. This is why it is often a good idea to have initial work on a project undertaken as part of initial planning. Having a high profile for a project that extends for a long elapsed time leads to greater risk.

- **Focus on both the details of the project and the wider view.**

  Project management involves concentrating on the detail of the project as well as having a wider perspective on what is going on and why. A good project manager should be capable of doing both and not focusing too much effort on one of the two extremes.

- **Align yourself with projects that are directed at key business processes.**

  Everyone wants to work on projects that produce tangible results and benefits. The types of projects that accomplish these things are those which improve standard business processes. It gets murkier when you stray into work that might be "nice to have." The problem is the absence of a pressing need for the project. Projects that do not link to major business processes may drift and be irrelevant. They have a much higher likelihood of failure. People won't want to participate in these projects because they do not see them as mainstream. Funding will be difficult to obtain because of competition with other projects.

- **Be aware of the impact projects have on the organization and direct the change and impact, rather than having it direct you.**

  Projects create their own dynamics. The end product of a project can transform an organization. An example is a reengineering project. However, success or failure can change the balance of power between managers of line organizations and upper management. A failing project manager is less likely to do well when he or she returns to the line organization.

- **Insist that all significant projects involve people from multiple departments.**

  Carrying out a project within and by one department may yield few significant results. If you undertake a project within a department, the visibility of the project outside the department is very limited.

- **Align the purpose and scope of the plan.**

  When you present an idea for a project, you raise people's expectations. They form a mental image of what the project will do. As time passes, this may change and expectations may rise as progress is reported. Reinforce the original purpose and scope of the project at each progress or issue meeting.

- **Inject humor and a sense of being a team through the sharing of experiences.**

Remember or reminisce about project experiences to capture the humor and joy in a project. Projects are dynamic and charged with deadlines and issues, but they are fun. Thinking about past project experiences can also help to lighten the stress in current projects.

- **Keep projects as small as possible, regardless of organization size.**

Many failures can be traced to large organizations. Government agency construction and aerospace projects are two examples. Large companies that attempt to undertake massive reengineering and downsizing with little or no results constitute another example. Large companies do not have to produce large projects.

- **Gather the team together and share lessons learned at the end of the project.**

When the project is finished, the team disperses. The end product of the project is visible for years. What is left from the project itself? To instill a sense of perspective and reinforce the experience for the entire team, share lessons learned.

- **Make sure that project success should support the organization.**

Project success does not equate to organization success. Even if a project is successful, the end product and the project itself can tear the organization apart. This occurred in one large aircraft manufacturing firm which designed and built a plane for the short haul market. The plane was a huge success. However, since it was not priced right, the firm lost money for each plane sold.

- **Know who gains and who loses from the success of the project.**

In any project you can identify potential winners and losers. If the project succeeds, the backer of the project gains. If it fails, the reverse may occur. The goal is to have the project appeal to the self-interest of many managers. This will increase their stake in the support of the project.

- **Enforce accountability.**

Lack of accountability for poor project leadership can increase failure rates. Sometimes a poor manager is transferred to another project. No one acknowledges that management problems existed. The manager carries the problems to the new position. If you enforce accountability, at least others will have some sense that failure or problems occurred.

- **Learn from past projects.**

People finish a project and immediately management throws them at the next project. Why not take the time to assess lessons learned? The answer is that no approved way to charge for time spent on this exists. In reality, assessment time should be built into the old project. If that is not possible because the project budget has been closed out, then build this assessment into the new project.

STATUS CHECK

This section of each chapter tests your application of the concepts in the chapter.

- Does the project you are working on have a clear objective and scope? Have these changed since the project was started?
- Is there an established process in your organization for setting up and running projects?
- For a project that was less than successful, was an effort made to develop lessons learned and apply these to other projects?
- Are your projects managed in a traditional way—project administration and management are handled by the project manager—or are they based on collaborative project management where duties and information are shared with the team?

ACTION ITEMS

Here is a set of tangible steps you can take to implement the materials of the chapter:

1. Projects and project management have been discussed. A first suggestion is that you assess the projects around you by reviewing the following statements. Answer each on a scale of 1 to 5, in which 1 means that you strongly disagree with the statement, 2 means that you disagree with the statement, 3 means that the statement is sometimes true, 4 means that you agree with the statement, and 5 means that you strongly agree with the statement.

   - Our company allows project leaders to use their own methods and tools in doing their projects.
   - Our projects take many good resources from the line organizations.
   - We do not have many tools or defined methods in general.
   - There is no organized training support for the tools and methods that we have.
   - We do not collect lessons learned from our past projects.
   - We do not have standardized project templates for projects. Each project starts from scratch.

   Use your response to assess the current problems with project management in your organization.

2. Define a simple project for yourself. Develop its objectives along the lines of what was discussed in the chapter. Next, define the potential scope of the project. Develop several alternative sets of objectives and scope statements.

   - How does the variation in scope impact the objectives?
   - Can you widen the scope without impacting the objectives?

3. For the project you defined in number 2, identify several issues that you are likely to face in terms of getting resources and the schedule for the work. Do this first for the narrowest scope and objectives. Then widen the scope and objectives. What new issues do you add? You will be working with this project in the action items of the following chapters.

4. Assuming that your organization has multiple simultaneous projects, how does the organization cope with multiple project management? More specifically:

—Is there an overall project summary?

—How are resources share and managed?

—How are issues that cross multiple projects identified and resolved?

—How is the mixture of project and non-project work managed?

CD-ROM ITEMS

This is a listing of the CD-ROM items that supplement chapter material. For Chapter 1, use the following:

01-01   Current Projects

Use this to analyze your current projects.

01-02   Dimensions of Project Management

Employ this to assess your current project management process.

01-03   Potential Problems with Projects

Use this checklist to identify issues in your current projects.

01-04   Project Evaluation

This file is useful to analyze a specific problem.

CHAPTER 2

DEVELOPING THE PROJECT PLAN

CONTENTS

2

DEVELOPING THE PROJECT PLAN

INTRODUCTION

You are now ready to begin to develop the plan. Traditionally, you would have started by making lists of tasks and resources. You might be instructed on the best way to enter the data into a project management software program.

Take a different approach. Answer some basic questions first:

- What is the purpose of the project? Answer this from different perspectives, including the perspectives of management, the business process, technology, and customers or suppliers.

- What is the scope of the project? What is not included within the project? What are the boundaries?

- What are the tangible benefits from completing the project? Again, answer this from different points of view. Ask the reverse question. If the project were not done or if it failed, what would be the impact?

PURPOSE AND SCOPE

The purpose of this chapter is to guide you in how to lay out the project. Studies and experience show that many failures can be traced back to this early period in the project. Had more analysis been performed, failure could have been averted.

Here are some examples where failure to plan caused failure of the project:

- A tank was designed by the military. It went into initial production. It was later discovered that the tank was too wide for most of the bridges in Europe.

- In California, the state transportation department attempted to create a high occupancy lane from the west side of Los Angeles over steep hills to the San Fernando Valley. Only the week before it was to open a test was performed. A bus was sent up the hill. By the time it reached the crest of the hill, its top speed was just 25 miles per hour. The project was halted.

- The Aswan Dam in Egypt was built to provide power and water control for the Nile. It was constructed in the tropical and desert areas of Africa. After construction it was discovered that evaporation and leakage to underground, inaccessible reservoirs drained much of the water. For years it was too expensive to route the power north.

The scope in this chapter begins with the initial concept of a project and ends with the completed detailed project plan.

APPROACH

Start to work on the project plan by defining the objectives and scope of the project. Progress through defining the players involved in the project, the schedule, the budget, the methods and tools, the team, and finally the plan.

Step 1: Determine the Project Concept

Construction projects may seem to have obvious objectives and scope, but they typically have political objectives as well as construction objectives. When a city builds a park, it is for the use of its citizens. However, it is also an opportunity to name the park and give someone recognition. How many parks do you know of that are not identified by name? If you define a narrow set of objectives and scope, you risk having a completed project with little impact. If you define it too broadly, you risk failure and noncompletion. Scope generally relates to complexity—the wider the scope, the greater the complexity.

Consider developing the project concept prior to developing a detailed plan. The ingredients of the project concept are as follows:

- Purpose of the project
- Scope of the project
- Benefits of the project
- General roles of the project—which organizations are going to do what
- Basic issues that the project may face. Examples are resource-oriented issues (getting people or equipment), political issues, and issues of participation if organizations are very busy.

Why do all of this now? Because you want everyone involved to have the same common vision of what the project is to accomplish. If you don't define the vision early, there will be more problems later. Your project, for example, will be more prone to scope creep, which refers to a gradual expansion of the project with no additional time or resources.

How do you evaluate specific objectives and scope? Here are some tests you can use:

- Do the objectives and scope fit with the organization? Are the purpose and scope aligned with each other?
- Are the objectives too broad or too focused?
- Are potential resources available for the objectives and scope you have defined? Have you already defined a project that is not feasible?
- Where are the areas of risk—both technical and managerial? This ties in with the issues.
- Are the benefits reasonable given the purpose and scope? In information technology, if the purpose is to install a system and the benefits are to increase productivity, you are not likely to achieve benefits because the purpose is too narrow. It does not include the business process improvement after the system has been installed. Is it any wonder that many IT projects are completed successfully without tangible benefits? The process was not changed.

Based on the scope of the project, you can decide how to divide a large project into subprojects. The more subprojects you create, the simpler the project, but the more complex are the interfaces between the subprojects. It's a trade-off. Several ways to divide the project are defined below. Each of these has pluses and minuses.

- **By organization**   This ensures greater accountability for each subproject but may make coordination between the subprojects a nightmare. This is typical when installing large financial software systems. This was also typical in Roman times when the military had strong leadership.

- **By project leader**   This approach starts with the people and divides up the work according to their skills and experience. This is a good approach with the right people. However, if you have gaps with no leader or if a leader moves on, you could be in trouble.

- **By function**   Here you would partition the project into parts that apply to specific functional activities. An example is a construction project, which has electrical, plumbing, and carpentry subprojects.

- **By geography**   This is the historic approach to managing large projects where authority was delegated to specific regions. It was so with the railroads of western America. Today, with the Internet and rapid communications, this approach may not be necessary.

- **By time**   This is also a traditional approach. You divide a project into phases. Each phase follows another. Each phase is a subproject. The problem here is that it forces the project to be sequential. It is difficult to establish a parallel effort if only one or two phases are active at one time.

- **By interface**   Here you begin by assuming that risk lies in interfaces between parts of the project. Therefore, you organize the project to concentrate on interfaces at the start.

A typical set up for failure is to divide the project by line organization. The line organizations will not get along. The issues and problems will not be resolved at the interfaces between organizations. The project will likely fail even if the parts in each organization succeed, which is doubly frustrating.

Having divided the project into subprojects, you are now ready to evaluate these. Here are some suggested questions to ask. Ask the same questions if you want to attach the new project to an existing project.

- How will you gather data across the subprojects to get a sense of what is going on overall?
- How will you identify issues that cross multiple subprojects and get them resolved?
- Is the risk spread among the various subprojects, or does it fall in one subproject?
- Are some subprojects too small to be viable and likely to fall within another subproject?
- Have you ensured accountability? Or, have you set up a situation in which project leaders from different subprojects may blame each other?
- Can resources be moved and shared between subprojects?

Example: Transportation

Creative railroad line builders realized that they needed to incorporate several subprojects into the building of the railroads in order to be successful. They sold land, established irrigation, and built towns, all to establish a market for the railroad. They focused heavily on developing the land; then they constructed railroads. This worked. In contrast, other companies needed the railroad to transport their own products. Typical here were coal companies that employed the railroad to reduce the price of hauling coal. Less subprojects were involved. The reduced costs of transportation and increased sales more than made up for the cost of the railroad without these companies expanding into land development. The key was appropriate planning at the initial stages of the projects.

Step 2: Assess the Project in the Business, Technology, Industry, and Organization Environment

Here is a list of factors that might impact a project and how that might happen:

- Technology
    - —New tools are available to make products
    - —Technology can be used to test quality
    - —Improved technology is available for sales and distribution of products
- Competition
    - —Ideas or concepts can be gathered from firms in other industries.
    - —Services or products are improved by competition.
    - —Your target market is invaded by the competition.
- Government regulation
    - —Increased government reporting is required.
    - —Regulations impact your subcontractors.
    - —Methods or tolls are regulated or prohibited in the project
- Politics
    - —A change of the party in power results in priority changes
    - —The project is canceled or redirected.
    - —The project is accelerated.
- Cross-impact examples
    - —Technology makes different products more competitive with yours.
    - —Government regulation restrains your competition, thereby reducing the need for your project.

How should you employ this list? First, determine which items can be employed in the project. Second, try to determine where risks lie up front, before the project is started. Third, use the list to validate your objectives and scope. Consider what items in the figure could disrupt the project if changes occurred.

Step 3: Develop a Strategy for the Project

A project strategy will be your approach for attaining the objectives within the scope and the environment. The strategy provides focus for the "how" of the project, just as the project concept provides the vision and focus for management. Let's return to the railroad example. For the transcontinental railroad in the United States the approach was to build from both coasts toward the middle. In other railroad projects, different teams worked on different sections of long stretches of the route at the same time. In one railroad in South America, more than 700 miles were under construction at the same time.

A strategy must address all parts of the scope. Thus, if the scope includes political factors, you must have a political strategy—even if you don't advertise it. A political strategy is an approach for dealing with potential problems, for advertising and marketing the project, and for getting support. You may have a stated strategy and several unstated strategies. For example, in a reengineering project the stated strategy might be to improve process and design and implement a new computer system at the same time. The unstated strategies might be to rightsize and restructure the organization.

What should your strategy address?

- How you will organize the project
- How you will select the project leader and team
- What will be the role of the team in project management
- How you will manage risk and address issues

In order to develop a strategy, define your approach for each of the above items. Consider political, organizational, and technological factors to refine the strategy. Define several alternatives for each, then evaluate the alternatives with the following questions:

- What is the least expensive strategy? The most expensive strategy?
- What is the strategy that will produce results most quickly?
- What is the strategy that is "politically correct"?
- What is the strategy that can minimize risk?

The more thinking and analysis that you do here, the clearer the strategy will be.

Step 4: Identify the Major Milestones and Initial Schedule

A good engineer and project manager in the 1800s could visualize the project plan based on experience. Without computers and extensive paper files, he had to keep a lot of information in memory. He had to be able to visualize the following initial milestones: route determined, survey completed, land and right-of-way obtained; materials and people procured; and work started.

Milestones

A milestone in the plan is a task that has no length or duration. It must be capable of being evaluated or tested to see if it has been achieved. For example, the railroad company would have as a milestone that the track is ready to be laid. This would mean that it is apparent that all the preparatory work of clearing the ground has been accomplished, the track material has arrived, and all the necessary tools are at hand.

Draw up at least 10 to 20 milestones for each subproject.

Next, logically relate the milestones between the subprojects in terms of dependencies. If you have a dangling milestone that you know should relate to another subproject, you are probably missing a milestone. You will likely need to add some milestones.

Next, take a piece of paper and lay it out sideways (landscape). The long side of the paper is the timeline and the short side is for the subprojects. Draw a horizontal line for each subproject. Put the milestones on each line for the corresponding subproject. Draw lines between the milestones to show dependencies. A general example is shown in Figure 2.1.

Initial Schedule

Make several copies of this chart. On one, start backward from the target date for completing the overall work. This will tell you when

---

Figure 2.1: General Example of Subprojects and Timeline

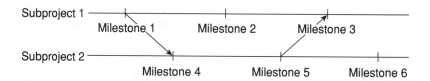

things will have to start so as not to delay the end date. On another chart, start at the beginning and estimate as you go. On a third, take a colored marking pen and highlight the milestones for which you perceive risk in their underlying tasks. Which milestones are these? They are the ones in which you have the least experience, the ones in which you really don't know how the work will be done, or the ones in which problems have occurred in the past with similar projects. Define the schedule based on these risky tasks.

Put all the charts side by side and create a fourth chart that reconciles these three. You can use the computer to do this. However, using manual methods gives you more flexibility. Such a technique was employed in the construction of several railroads.

Step 5: Define the Initial Budget, Using the Milestones

Think about what resources will be necessary to achieve each milestone. List 4 or 5 key resources for each milestone. Now develop an initial budget by milestone for each subproject. Always develop your initial budget bottom up. If you do it top down, you will miss part of the plan and resources. After you have completed this, you can estimate overhead and other resources as a group. Include facilities, supplies, and equipment as well as personnel.

Here are some common mistakes that people make:

- Failure to adequately consider downtime waiting or rework.
- Failure to allow for some change of scope in the project.
- Failure to consider potential additional tasks and work, resulting in underestimation.

In doing your budgeting, plan on holding onto resources only as long as needed. In modern projects you seek to release resources as soon as possible to reduce costs. For personnel costs, use a high average cost for an employee. Add the overhead cost in as well. For facilities and equipment include the set up time and tear down time, if applicable. If you are involved in new technology or methods, include training costs along with potential travel costs.

Many projects have two budgets. Since the project plan cannot detail all of the costs, use a spreadsheet to develop the realistic budget. The second budget will reflect the work performed by the resources. When you perform budget analysis, you often will extract this second budget from the project management software and then incorporate it into the spreadsheet.

Step 6: Identify Which Groups and Organizations Will Be Involved

Get out an organization chart and a piece of paper. Draw two vertical lines about $1/3$ and $2/3$ of the way across the page. Write down the organizations on the left and their roles in the project in the middle. Write down how important their involvement will be in the project in the right column. Define a role for almost every organization. Any that you omit can come back to haunt you. Include outsiders as well.

For the railroads, you would include in the left column the lobbyists and the agents for selling land and attracting settlers. By the way, railroads led the way in separating finance and accounting from the day-to-day operation of the railroad. They also began modern large scale financing.

Step 7: Determine the Methods and Tools to Be Employed in the Project

In many projects people begin the work and then select tools and methods as they go. The homeowner, doing repairs around the house on the weekend, is an example of a person who starts a project and ends up running to the hardware or lumber store many times. Use two sets of methods and tools, one for the actual work and the other for the project. This is important since it validates the schedule and budget estimates. It also forces you to identify where you have holes and gaps at the start.

Step 8: Identify the Areas of Risk and Associate Them with Milestones and Tasks

Risk can arise from a number of factors. You may not know what is needed to produce the milestone. Perhaps the organization has never carried out this sort of work before. You may have to rely on unknown internal staff or external contractors to do the work. The milestone may be more complex than first perceived. Another factor is lack of definition of the tasks. Sometimes you just don't know what is an acceptable milestone. New technology and lack of direct control are other sources of risk.

Label the milestones that obviously have substantial risk. Now attempt to define some additional, more detailed milestones within each of these. If you can divide a major milestone into smaller milestones, you might be able to reduce the risk, or at least isolate the risk to smaller milestones.

From the work so far, you can start with the list of issues that may impact the project as defined in the project concept. Use the list of issues at the end of the book in the last chapter as a start. With the list in place, scan the tasks in the plan. Identify any tasks to which an issue pertains. Label the task with the issue. You now know these tasks have risk. Go through the list of tasks. You will find some that have risk. For each task that has risk, go to the list of issues and find the associated issue. There are two situations you will face in addition to matches. First, there will be issues that have no tasks. However, these issues are valid. Therefore, tasks are missing. There may be tasks that are risky but have no associated issues. This means that you have missing issues.

This step is valuable in validating the tasks so that you can feel that your plan is complete. It also helps you to identify tasks that have risk and the source of the risk.

Step 9: Refine the Schedule and Budget

Refine the estimates of budget and schedule. If they don't change, you are either a good estimator, or you have missed something.

Step 10: Identify the Project Manager or Leader

Think about candidates for managers for the project and for the subprojects. Identify several alternatives, if possible. Think of availability and the potential of not having them throughout the project. You will need a backup plan for a project leader as well. This subject will be discussed in detail in the next chapter.

Step 11: Identify and Establish the Project Team

This is addressed in Chapter 4. In this initial stage of the project, identify a few key people that you need for the core of the project team. Decide how many team members you will need, then determine who you will choose to be the other team members by evaluating skills and knowledge. You can use the results of your budgeting and scheduling effort to help you here.

Step 12: Develop the Detailed Project Plan

You now have the knowledge to develop the tasks of the plan.

1. For each subproject enter the milestones and the resources that you identified. These are in two separate lists.

2. Now define the tasks that lead up to each milestone. See a detailed explanation following. You now have a work breakdown structure with a list of tasks. Go down to the detailed level for the initial phase of the project. Keep the tasks at a more summary level for later parts of the project (several months out to the end of the project).

3. Establish dependencies between tasks.

4. Assign up to 4 or 5 resources per task. Don't attempt to be complete in terms of all possible resources. Each resource should represent a job function or type of facility or equipment. Don't attempt to name people or individual components at this point.

If you are using project management software, you now have a template.

5. Estimate the duration of each task and set the start date of the project (if not already determined by the dependencies). Make sure that the tasks assigned to be undertaken "as soon as possible." This will provide the greatest flexibility. The schedule will unfold.

6. Assign the quantity of each resource for the tasks.

7. Analyze the schedule and make changes by changing duration, dependencies, resources, and starting dates.

Save the file at the conclusion of each step. In this way you can go back and begin again if needed. Do not fill in the detail later in the project. You are going to have the project team members do this later so that they will be more involved and committed to the project. This method was employed on the railroads when the site engineer set the detailed tasks.

Major disagreements or issues may surface during these steps. The project may not get off the ground. This is not a cause for despair. It is better to know at this stage that the project won't work than to find out when the project is underway and money and resources have been consumed.

SETTING UP TASKS

Here are some suggestions for setting up tasks so that they will be easy to use and work with.

- Keep the task description simple—less than 30 characters.

- If the task name is compound or complex, split the task.

- Start each task with an action verb.

- Use a field in the project database for responsibility for the task. This is different from resources since the person responsible may direct resources in the performance of the work.

- Each detailed task should be from 2 to 10 days long (shorter tasks mean too much detail; longer tasks mean that the task is too general and cannot easily be monitored).

- Use standard abbreviations wherever possible (e.g., Dev for Develop).

- Number all tasks in an outline form (e.g., task 1100 is the first task under task 1000).

- Establish categories of resources (personnel, equipment, facilities, etc.).

- Try to avoid using the individual names of people—put a job title in abbreviated form instead.

- Keep resource names to less than 10 characters.

- Use a field in the software to indicate which tasks have substantial risk (e.g., a flag field that is either yes or no in value).

- Use task outlining and indenting.

- Group the tasks with appropriate milestones.

- Label milestones as such (e.g., "M: Foundation completed" and "M:" indicates a milestone).

- Use a field to put in the name of the person or organization accountable for the task.

Let's amplify some of these suggestions. The reason for short names and abbreviations is to have more readable GANTT charts and reports. The numbering of tasks makes it easier to follow later when tasks are added or changed. Resource categories are useful in filtering and reporting by type or category. Most software has the capability of customized fields in the database. The flag to label tasks as risky allows you to extract only these tasks for evaluation and review. You can also define all paths that pass through these tasks. Define any of these paths to be a management critical path. The reason for not having names as assigned tasks is to provide flexibility if assignment changes for individuals, but the same person is performing the task. The reason for the use of a separate field for accountability is to distinguish between who is accountable and what resources are required. These are not the same.

A later chapter considers presentations to management and communications. Each review has a single purpose. In the first review, give others just the scope of the project and the lists of tasks and resources. This gets them accustomed to the terms. In a later review, discuss the completeness and focus of the scope of the project.

During the next review, show the dependencies with the tasks, without duration or dates in the schedule as yet. The purpose of this review is to ensure that the relationships are correct.. The third review is to show the tasks, dependencies, and assigned resources so that people can agree on what the key resources are for each task. Leave some blank to generate interest.

If you wait for everyone to participate and join in the steps presented, you are likely to be disappointed. Take the view that the developer of the initial plan and the project manager should be proactive. Take a crack at carrying out all 12 steps. You will then have a working plan that consists of lists of items, the budget, and the project plan. This will accomplish several things. First, you will gain confidence in the schedule. Second, you will be more focused when you ask for reviews or input on the plan. Third, you will be more organized. Therefore, you will be more likely to be successful.

GETTING STARTED

It is sometimes a challenge to get started, even with the steps and suggestions provided. If this is your first experience in project management, start by doing some of the steps for an existing project where information is available. This will help you get familiar with the pattern of this analysis.

PLANNING THE TRANSITION FROM THE PROJECT TEAM TO A LINE ORGANIZATION

Even though this is the start of the project, plan now for the eventual transition of the project from the project team to an operational line organization. This is important, since a project may be successful up to the transition and then fail.

Here are some things to include in the planning:

- Identify the organization that will be responsible for the results of the project.
- Work with the organization to determine several people who will be responsible for day-to-day operation.
- Plan a limited role for these individuals in the project before the transition to get them committed and involved in the project.

If the scope of the effort is small, go through all the steps anyway. Later, if the project expands or if problems arise, you will already have at hand formal project management methods and tools.

PROJECT DOCUMENTATION

Obviously, documentation depends on the size and complexity of the project. You can justify the time spent on this documentation on the grounds of managing risk and for marketing.

The minimum needed for a very small project is a task list. No formal schedule is needed.

Here is a list of items recommended for complex and large projects:

- A project plan for the overall project
- Detailed project plans for each subproject
- A list of initial known issues for the project
- A description of interfaces between subprojects
- A description of the roles of organizations in the project

Remember that most of these do not live on and are not maintained. The key exception is the project plan.

A basic guideline is to adopt a zero-based approach. That is, start with no documentation assumed. Then justify each document that is needed. This will help ensure that the project will generate the minimum amount of documentation.

DO YOU HAVE A WINNING PLAN?

After you have developed the working plan or the initial version of the plan, evaluate it yourself. Be your own worst critic. Put the plan aside for several days and work on other activities. When you reopen the plan, ask the following questions:

- Are the objectives and scope consistent?
- Is the scope reflected in the range of tasks?
- Is the strategy borne out in the tasks?
- Have you identified the areas of risk?
- Have you defined the key resources?
- Have you associated tasks that carry risk with the list of issues?

- If you were assigned the job of attacking the plan, what would you see as the major weaknesses?

EXAMPLES

The Railroad Example

In general, project planning in the railroad industry started in a very primitive way. Within less than a decade, standardized approaches began to evolve, due to the need for estimating schedules and budgets accurately. Even then it took years to increase the accuracy and refine the approach. In early years, the project planning work often was performed by a few key engineers. As time went by, more people became involved. Other industries connected with the railroad also became involved and had to become better organized.

Modern Examples

In the manufacturing example, the detailed project plan was developed first without the earlier eleven steps. This presented problems in marketing the plan since management was not involved in the development of the plan and sometimes resented the plan being thrust upon them. Ideally, plans should be developed so that people will become interested and committed.

A better approach would have been to develop parts of the plan, get it reviewed, revise it with feedback, and continue with the steps. In the end this would have saved time and reduced misunderstandings.

The project manager at the consumer product company had never developed a plan in this area. He decided to use several acceptable plans as a basis for building a template. This worked because he began from a known base.

GUIDELINES

- **Build a plan with great detail on the near-term tasks but less detail for tasks that are further out in the future.**

This will allow for flexibility in working with those future tasks. Also, project team members will then have the opportunity to participate and fill in details as the project progresses. On the other hand, if you build a detailed schedule for the entire plan, this schedule will have to be revised often, based on actual results. It may be too restrictive and may lead to disruption later.

- **Take a large project and divide it into phases.**

  In a given phase, identify the major tasks and milestones to see if you could increase effort and move tasks up in the schedule.

- **Consider how much time you have to spend on updating the plan when you design the plan.**

  If you design a complex and detailed plan, you will have to spend more time updating the plan. For example, if the lowest level of detail is two to three days, you can update the schedule twice a week. If you go down to tasks of one day or less, you may have to update the project daily.

- **Look at the project's external appearance to learn about the past and present of a project.**

  Examine the project from the outside. What are the perceptions of managers outside of the project? How has the planned budget and schedule tracked against the actual? Have people left the project? Why?

- **Use a chart to create a picture of the project.**

  A project can be thought of as an eight-dimensional figure— project plan/schedule, project manager, management, user, staff, purpose, scope, and methods/tools.

  Construct a bar chart or a radar chart with each bar or line signifying one dimension. You can use this to compare different projects in each of the eight dimensions. Charts such as this give a picture of a project without plunging into detail. You can also consider alternative purposes, scopes, etc. for the same project.

- **Start projects based on a fiscal year to avoid resource conflicts.**

If your project begins at the start of a fiscal year, the project has to compete with other projects for attention and resources. If possible, begin three months after the start of the year with funding that was approved for the start of the year.

- **Avoid getting sidetracked by the process of a project.**

  When you consider the relative hardships of long projects and large projects, the more difficult is the long project. A long project can transform the project team into a pseudo line organization. Watch that the team does not get caught up in the process of the project as opposed to the actual work.

- **Remain sensitive to the environment throughout the project.**

  The environment of a project was covered in the second step. The project not only must be planned with these factors in mind but also assessed during the life of the project for changes in the environment.

- **Understand what not to do in a project.**

  Start the project with tasks that you know have to be performed. If you start adding tasks that might be needed, you could escalate the project cost and work. You will divert attention from the important tasks.

- **Hold one person accountable for each detailed task.**

  If you have to identify two people for a task, then split the task into two parts.

- **Minimize documentation.**

  In a project you can devote your time to doing either project work or to administrative tasks. Documentation is an administrative task. It may be necessary to produce the documentation. However, working on documentation may mean spending less time on the project itself.

- **Analyze risks at the start of a project.**

  This reinforces doing extensive analysis and planning prior to and at the start of projects. If you understand the risk areas, then you can give them proper attention.

- **Choose longer elapsed time over greater effort.**

  If you ever have an opportunity to choose between more time and more people, choose the time. Also, introducing more resources will likely impede the project. If you are offered more resources, do not accept these at face value. You may discover hidden costs in the politics of procuring the resources.

STATUS CHECK

- Does your firm follow an established sequence of steps in developing project plans? If so, are these clearly distributed and supported by training?

- How are small projects handled differently from large projects in your company? What happens to projects that are in the middle?

- If you were to develop a new project plan, what guidance, templates, and other support are offered in your organization?

- In what areas of project management are your company's greatest strengths and in what areas are the greatest weaknesses?

ACTION ITEMS

1. A first possible step is to evaluate a plan with which you are involved or familiar. Use the guidelines discussed earlier.

2. After doing this evaluation, sit back and try to determine what could have been done at the start of the project to head off the problems.

3. Now look over several projects and attempt to define the objectives, scope, and strategies for the projects. Create a table in which the rows are the projects and the columns are objectives, scope, and strategies.

4. For a simple project, develop a template using the following steps:

   - Identify 10 to 15 highest level summary tasks.

- Identify the same number of major milestones. Put these into a task list. You now have a very high-level work breakdown structure.

- Identify dependencies between all tasks and milestones. Note any cases in which you are having trouble deciding if there is a dependency. Later, you will be able to determine such dependencies when you create more detailed tasks.

- Now, for each major task, write down the detailed tasks. Put these under the summary tasks.

- Identify five key resources for the project. Assign these to the detailed tasks.

- You have now created a project template.

5. Take the template you just created and add more detailed tasks for the work to be done in the next month. Define the duration of all detailed tasks. Determine the starting dates for all detailed tasks that do not have predecessor tasks. You have now created a schedule. Flesh out the tasks one month in advance on a regular basis.

CD-ROM ITEMS

02-01   Development of the Plan

02-02   Project Concept and Initial Planning

02-03   Powerpoint 95 Project Concept Presentation

This is a program to be used for presentation of the project concept to management.

02-04   Custom Project Template

02-05   Project Template for New Product Debelopment

This file is an examples of a plan for vertical markets.

These Access databases and Excel spreadsheets are for issues and lessons learned:

02-06	Access 97 Database—Issues
02-07	Access 97 Database—Lessons Learned
02-08	Issues Database (Excell)
02-09	Lessons Learned Database (Excel)
02-10	Software Project Template

CHAPTER 3

THE PROJECT MANAGER

CONTENTS

3

THE PROJECT MANAGER

INTRODUCTION

While the term *project* is less than 100 years old, projects and project managers have existed since the construction of the pyramids of Egypt. At the heart of the role of project manager is the focus on achieving the objectives of the project. A look at thousands of past documented projects reveals key attributes of a project manager. Many successful project leaders share the following:

- Problem-solving capability to identify and to resolve issues associated with the project;
- Steadfastness to see the project and tasks through to completion;
- Ability to work successfully with the project team, management, and other employees and outsiders.

Notice what is *not* on the list: being clever or being a technical genius. Notice also that the characteristics described above can be developed. Project leaders are not born, they are made. Most successful project leaders grew into the role by necessity. Even if you are not a project leader, you can use the material in this chapter to build your skills and prepare for the time when you become a project leader.

PURPOSE AND SCOPE

The purpose of this chapter is to explore the most significant parts of a project manager's duties. Administrative responsibilities will be examined. However, your success will rest on your ability to deal with people, to address issues, and to analyze and demonstrate leadership, so these attributes will be emphasized.

The chapter covers not only principal activities, but also situations with which you are likely to be confronted. The scope extends from the start of a project to the completion. Sometimes you are thrown into the middle of an ongoing project. This situation will also be covered.

APPROACH

One might say that a project manager's role is to implement and complete the project. However, this is the narrow version of the scope of a project. A wider view is to work towards leaving a project management process in place for people to use after the project is completed.

People sometimes fail as project managers because they never clearly define their role. A project manager who embraces the role in a broader sense can create a more aggressive approach to the entire project, while a project manager with a mindset of a narrow role may be defensive and weaken the project.

PROJECT MANAGER SELECTION

Unfortunately, many times the choice of a project manager is made on an ad hoc, spur-of-the-moment basis. Often, the people are selected based on availability and a general experience fit with the project. In projects that are routine and of low risk, this will probably work. This fails, however, if the project manager underestimates or misunderstands the risk and exposure involved in a project.

What is a better approach for selecting a project manager? Here is an organized method:

- Have managers maintain a list of people who appear to be potential project managers and update this annually. Scour the organization to find these people. The benefit of this list is that it gives management a reference point from which to begin.

- When a new project appears or is being considered, round up all other project ideas that are likely to turn into projects in the next three to six months. Identify the degree and source of risk in each project. Sources include organization coordination, systems and technology, and external organizations. Construct a table where the rows are the projects and the columns are the areas of risk. In the table, rank each project according to the specific area of risk on a scale of one to five where one is low, or no risk, and five is very high risk. This table indicates what skills you need for each project to minimize risk.

- Take the list of project leader candidates and add the names of project managers who will be available during the period. Construct a second table in which the rows are people and the columns are areas of risk. Enter a one to five in the table, based on the degree to which the person can deal well with that type of risk. This shows the most suitable areas for each candidate to handle.

- Now you can compose a third table of project leader candidates (rows) and projects (columns). The entry is the extent to which each person is suited to the specific project. Note that you cannot just put the previous two tables together. The table here reflects knowledge and familiarity with the project as well as with handling risk.

Errors in this approach come mainly from misunderstanding the project and its risks, not from misunderstanding the people. Often, there is insufficient analysis of the project before a project leader is selected. Carry out the pre-project activity as presented in Chapter 2 before you use this method of selecting a project manager.

PROJECT MANAGER RESPONSIBILITIES

Consider the table in Figure 3.1. The columns are the phases of a project from start-up to after project completion. The three rows are for major duties, administrative duties, and background duties. Major duties are where you should spend the most time. Administrative duties are overhead tasks that are necessary. Background duties are things that you can do to help yourself be a better project manager. At any given time, have several of these activities in process with one getting primary attention. Give primary attention to the major duties rather than to administration. Also, gradually work at improving yourself through the background duties.

Figure 3.1: Duties of a Project Manager

Type\Project Phase	Start of Project	During Project	End of Project
Major Duties	• Define objectives and scope • Define the project plan • Market the project	• Identify and address issues • Monitor the actual work • Make decisions and do marketing • Interact with line managers and team • Review the project	• Construct lessons learned • Support placing project members • Find a new project
Administrative Duties	• Set up the project files	• Update trhe project status • Perform a budget vs. actual analysis • Revise schedules and budgets	• Clean up and store project files • Document lessons learned during the project • Follow up on any loose ends in the project • Create a final budget vs. actual analysis with assessments of variations
Background Duties	• Conduct casual marketing of the project and plan • Line up staff for future work later in the project	• Track what is happening on other projects that have interdependencies with yours • Exploit common ground, issues, and opportunities with other project leaders	• Build ties with other project leaders • Apply lessons learned

- **Major Duties at the Start of the Project**

  Define objectives and scope. This is a high priority. If you accept the objectives and scope you are given without analysis, you may find that they do not fit together. You typically have a chance at the start of the project to provide input on the objectives and scope. Use this time wisely.

  Define the project plan. Take the time to develop the project plan carefully. Typically, the first few attempts are not very flexible or complete. You will need time to think about risk and the impact of risk on the plan. Start early.

  Market the project. Even if the project has been approved, it has not been sold to everyone. What is in the project that appeals to the self-interest of the organizations and individuals involved? By doing marketing and sales, you force yourself to consider the project from points of view other than management's and your own.

- **Major Duties During the Project**

  Identify and address issues. Keep on this subject constantly. If you let up, you risk the entire project. Act as a constant problem-solver. Look for issues. Make sure that the issues that have been resolved do not resurface under a different guise.

  Monitor the actual work. Go out and actually see what is going on in the project. Do not take people's word for it even if they are good and truthful team members. By visiting them while they are working and showing an interest you also show that you care.

  Make decisions and do marketing. You have to do marketing to convince someone that the decision is needed, that the timing is important, and that the form and structure of the decision are correct.

  Interact with line managers and the team. Keep line managers informed of what is going on in the project and how their organization is contributing.

  Review the project. This is not a background task. Actively set aside time for analysis and perspective.

- **Major Duties at the End of the Project**

   Construct lessons learned. As the project winds down, develop a list of lessons learned with the tips provided. These demonstrate the added value of the project to the organization, as well as showing that you really do care about more than the single project.

   Support placing project members. Long before the people return to their line organizations or go on to other projects, help market them to ensure that they get positions using the strengths that you have observed during the project.

   Find a new project. Even if you are wildly successful, your next project will not often materialize automatically. Market yourself by volunteering for other work. Don't be stuck on one project. Show that you are interested in several projects.

- **Administrative Duties at the Start of the Project**

   Set up the project files. This means not just paper files, but also electronic files, templates, forms, and all of the support required for the project. If you take the time to do this with care at the start, you will save yourself grief and problems later.

- **Administrative Duties During the Project**

   Update the project status. Keep track of status of the project and keep management informed. Don't assume that if you tell one manager, other managers will be told. Inform people one-on-one of what is going on.

   Perform a budget vs. actual analysis. Get in the habit of routinely looking at the budget so that you constantly know the status of the budget.

   Revise schedules and budget.

- **Administrative Duties at the End of the Project**

   Document lessons learned during the project.

   Clean up and store the project files.

   Follow up on any loose ends in the project.

   Create a final budget vs. actual analysis with assessments of variations.

- **Background Duties at the Start of the Project**
  Conduct casual marketing of the project and plan.
  Line up staff for future work later in the project.

- **Background Duties During the Project**
  Track what is happening on other projects that have interdependencies with yours.
  Exploit common ground, issues, and opportunities with other project leaders.

- **Background Duties at the End of the Project**
  Build ties with other project leaders.
  Try to apply the lessons learned.

THE IMPACT OF TRENDS

The basic role of the project manager has not changed through the years. Some trends have made the work easier and some have made it harder.

Here are some trends that have affected the work of the project manager:

- Availability of software for electronic mail, groupware, and project management has eased some of the administrative and communications aspects of the project.

- Corporate downsizing and reorganization have made projects leaner and more accountable, making projects more challenging.

- The availability of new technology and project successes has increased management expectations.

- Improvements in technology and their business impact have increased pressures for projects to succeed.

- Resources have to be shared among projects and with non-project work, creating a coordination challenge.

- Fewer resources are dedicated to an individual project, producing a management challenge.

Overall, projects are more challenging than they were 20 years ago due to tight schedules and limited resources. The upside of this is that in many organizations you can more readily advance in the organization by succeeding with a project. The trend is moving toward more projects and, consequently, toward more project managers. Projects appeal to management in many organizations because of the accountability and visibility. As one manager said, "Projects can generate revenue; line organizations generate costs."

THE POSITIVE ASPECTS OF BEING A PROJECT MANAGER

Now that you have seen what a project manager does and how the role has evolved over time, you might ask, "Why would someone want to become a project manager?" It looks like a lot of work. It entails career risk. If you are successful, you might just end up doing one project after another. All of this is true. Yet, people still seek to become project leaders. Why is this so? What is the positive side of being a project manager? For younger people it is an opportunity to gain the recognition and attention of management. A project is visible. A project manager is visible. A project manager may have more opportunities for advancement than an employee stuck in a line organization.

Here are some additional reasons to become a project manager, beyond the potential of advancement:

- A project may be a chance to learn new skills and methods.
- Being a project leader may be your best shot at breaking into management.
- The role of project leader may be more challenging than your current position.
- Being a project leader will increase your range of contacts and personal network.

BECOMING A PROJECT MANAGER

If you are a project manager with a known track record, you can wait to be called. Otherwise, how do you get started? Consider the old-fashioned approach of volunteering. Even if you are turned down for a particular project, people will see that you have an interest in the company. Second, you have shown initiative. How many people around you are doing that? Third, you are alerting people to your interest in projects and project management.

What should you do before volunteering? Scout the project out. Determine what you could add to the project. Don't stress the negative things in the project that you could fix. Focus instead on the opportunities in the project and the strengths you could bring to the table. Next, determine what you would do with the duties of your current job. You definitely don't want to say that these can be left undone. Instead, show that you can piggyback the project tasks on top of your normal work, even if this means extra work.

What if you want to volunteer to replace a current project manager? Do not undermine the current manager. How do you propose your services? The best technique for taking over a project is to volunteer to address the outstanding, unresolved issues in the project. Never emphasize administrative or communication skills unless there is a known problem here. You want to show that you care and want to help.

When you make your proposal to management, use the words "volunteer" and "assistance," and ask them to think about it. To avoid pressuring them for a decision, indicate that you will check with them in a week or so to see what they might suggest.

SUCCESSFUL PROJECT MANAGERS

Here are some of the most common characteristics of successful project managers:

- They know what is going on in the project at any time.

  Be ready to answer any reasonable question about the project from anyone. This will show that you are in touch. It will show the team that you care.

- They work on communications with line managers.

  Stay in touch with the line managers who are responsible for your team members. This way, they will know about the contributions of their people to the project and will be less likely to remove them from the project.

- They are aware of the trade-off between the needs of the organization and the needs of the project.

  Many times both the project and organization have a common interest. However, sometimes a decision is made one way for the organization and another way for the project. When you press for a decision, point out this trade-off so that all involved can see how the decision will affect the project if they decide in favor of the organization.

- They can address resource allocation among multiple projects dynamically.

  A modern project manager must often compete for resources for the project on an ongoing basis. This is true even if the project is recognized as critical.

- They are able to evaluate and criticize themselves.

  Be your own worst critic. However, also pat yourself on the back when you succeed. That is part of the evaluation.

- They have a sense of humor.

  Look at the humorous side of projects. Consider how ridiculous all of the red tape and bureaucracy are. Dilbert cartoons often point out the absurdities of organization life and sometimes give a humorous view of project management.

- They work with project team members one-on-one to understand their needs and frustrations.

  This includes working with the people who are part-time players in the project as well as with your core team. It may be casual conversation away from the project.

  Treat the team members evenhandedly. This is difficult since certain people at any given time are more critical to the project than others. Also, some people in the project may be taken for granted. When the pressures of the project build up, this continues. The project manager is the one who can most easily get such people recognized in their own organizations or with management.

- They are always on the lookout for ways to improve the project and the environment of the project.

  Listen to the team members for their ideas. Solicit suggestions as to what could be done to improve the work. Don't mention budget or schedule. This will just increase pressure. Ask the question, "Do you have any thoughts on how the work might be accomplished in a better or easier way?"

- When decisions are made, they act immediately.

  Prior to the management decision, map out a plan of action assuming that the decision goes the way you anticipated. Be ready to act when the decision is made. This is not just to show action. You wanted that decision; you had said how important it was. If you sit on your hands after you have the decision, you may lose credibility.

- They become adept at the methods and tools.

  The purpose of this is to be self-sufficient. You do not want to have a critical deadline come up and then have the person on whom you depend on be unavailable. While you cannot be an expert, you should know enough about most tools to get by.

- They practice project fire drills in the planning.

  Be ready for emergencies. This will also help you deal with the unexpected. For example, you show up for work on Monday and a manager comes in and says that someone is being pulled off the team for a high-priority task. Or management asks you to determine whether and how the project could be accelerated. Think through these and similar scenarios and formulate plans, both to have the plans ready for possible use and to practice thinking through problem situations.

PROJECT MANAGERS WHO FAIL

Here are some common reasons for failure as a project manager:

- They avoid being involved in the actual work.

  This is a sure ticket for trouble. You risk losing touch with both the work and the team. Also, if you roll up your sleeves and do some of the work, the team will respect you more.

- They try to micromanage the project.

  This can irritate the team members. The project manager might cruise the area where the work is being done and direct people in the smallest task. People notice the project manager's presence and start to ask, "Doesn't this person trust us?" or "Why can't we do the work ourselves?" Instead, delegate tasks and then follow up on tasks that have risk.

- They attempt to deal with issues one at a time without analysis.

  As will be discussed in a later chapter, issues tend to link together. Some issues may continually resurface due to political factors. Sit down and analyze these issues and then attempt to resolve a group of them at a time.

- They leave project administration alone or delegate it.

  Administration is downplayed compared with other work, but you still should do the reporting and analysis of the project yourself. If you rely on others, you may not be able to answer questions when asked casually by managers in a meeting or in the hallway. Any hesitation may be viewed as a sign of weakness.

- They spend too much time schmoozing with upper management rather than spending time on the project.

  The more time you spend with management, the less time you have for the project. It is a zero-sum proposition. Balance your time. Also, do not wear out your welcome with management.

- They spend excessive team meeting time on status.

  Get status one-on-one before the meeting. Use the meeting time to address issues and opportunities.

- They become obsessed with how many tasks have been completed and the percentage of work completed towards a milestone.

If this occurs, you are becoming a bureaucrat. A milestone is only complete if you validate that the work is of high quality and fulfills its purpose. Percentage complete means nothing if you cannot validate it.

- They leave issues, especially old issues, unresolved.

  Issues left unresolved tend to fester and get worse. On the other hand, you might want to allow an issue to mature until you understand its ramifications.

- They make too many changes at once or minor changes too often to the project schedule.

  This can irritate team members because the project then appears to be adrift. Make several changes at one time with an umbrella reason for the changes. Then leave it alone for awhile.

- They become focused on either the tools of project management or the tools used by the project.

  Tools are often technology-based. They are impressive and captivating. Don't be sucked into this trap. Tools support a method. Pay attention to the method. Let others worry about the tools. Your main concern is whether the tools support the method and are being properly used, not the internal workings and features of the tools.

MEASURING YOUR PERFORMANCE AS A PROJECT MANAGER

How are you doing as the project manager? To make the most of your experience and come up with lessons learned, use this checklist to evaluate yourself:

Checklist for a Project Manager

- How much time are you spending in project administration vs. profit management? Is the balance good or does it need adjustment?

- What is the actual state of the work in the project right now? What are the problem areas?

- Do you communicate informally with upper management on the project? Do you need to communicate more often?

- List the key issues that remain unresolved? How long has each remained so? What is the age of the oldest unresolved issue?

- Is the project plan and schedule up-to-date? If not, what areas need to be brought up-to-date?

- Do you communicate with members of the project team one-on-one frequently enough?

- What is covered at meetings? Do the meetings last too long? Are the meetings rushed?

Your team members will benefit from periodically evaluating their own work on the project, also. Use the following checklist with your team members. Discuss with them areas that need to be improved, either with more effort on your part or on theirs.

Checklist for a Team Member

- Do you have an adequate picture of the overall project status?

- Have you taken care of issues assigned to you for analysis and resolution? Are you unsure about how to proceed with any of the assigned issues?

- How much time are you devoting to the project vs. other work? Is the balance good or does it need adjustment?

- Are you using the methods and tools of the project correctly and effectively? Do you need more training?

- Do you volunteer to do additional work?

- Have you had any problems getting along with the project leader?

- Do you communicate what is going on in the project with your line manager?

EXAMPLES

The Railroad Example

In the 1800s, several distinct types of project managers supported railroad construction. First was the general engineer who selected the route. This person literally walked and rode hundreds of miles over hills, through valleys, across rivers, and through rugged mountains to find a feasible route. An additional project manager was the head of the surveying team which followed the general engineer. As construction revved up, each area of construction had a project manager. The marketing of the land and the attraction of immigrants required still another type of project manager.

The railroads organized work along project lines because of the sheer scope and size of the operation. At one time the Pennsylvania Railroad had more employees than the entire postal service (the largest government agency at that time). The amount of capital required, the logistics, and the management of the men all pointed toward formal project management methods.

How did these project managers emerge? Many young people joined the railroad. Provided that they did a good job and did not get injured, many opportunities for advancement existed for them. It was possible for them to work their way up to being a project manager.

It was common for railroads to go bankrupt and then reemerge from bankruptcy. Thus, some people worked for a number of different railroads. Others worked for the same line but were frequently transferred, since that was the only path for promotion. In both cases, the growth of experience and lessons learned occurred. Since most of the procedures and communications were verbal, it was important that people were trained and self-sufficient. This was a prime reason for the extensive apprenticeship programs conducted by the railroads.

The railroads produced outstanding project managers. Several could design a route in one visit while riding a horse through an area. Then they would return to camp and estimate the cost of construction and draw up an accurate schedule. Railroads also had some of the best managers of promotions to entice immigrants from Europe. Logistics in many cases were superb. Rail, ties, spikes, and fish plates arrived on schedule for work. The requirements of the railroads on other industries were intense. Steel mills, agriculture, coal, and lumber all had to meet production requirements and quality standards. This created a need for project managers in these industries to oversee the goods being supplied to the railroad.

Modern Examples

Management in the manufacturing firm knew that the coordination of a project at ten international locations and headquarters required excellent human relations skills. They also knew that they wanted someone with experience and knowledge of operations. Ideally, they desired technical knowledge for networking. The final selection boiled down to three candidates. One was technically oriented and had carried out smaller network projects in corporations. A second had extensive experience in Asia, had rapport with surrounding offices, and was reasonably experienced within systems. The third had worked on non-technology projects with the divisions while lacking technical knowledge and systems experience. Fortunately, the managers met and discussed where the risk was greatest. They agreed that the organizational and human relations side of the project was much more complex and important than the technology behind the network, even though the company had not implemented such a network before. On this basis they chose from the three candidates.

The purpose of the project at the construction firm was to implement an overall project management process. The various project plans and projects were to be structured into common templates to allow analysis and gathering of lessons learned. There were several candidates for the project manager position. In fact, the eventual manager did not get the job at the start. His expertise was in carrying out several successful projects. He had not worked within the corporate project scheduler group. Management initially chose the head of the project schedulers. This seemed like the natural choice since this candidate was in charge of current scheduling. It was a big mistake and bordered on a disaster. The project started with the assumption that the scheduler role would not change and that the schedulers would increase their power over the project managers. The change occurred when a band of project managers stopped the project and went to management to protest. Management asked them for input. Because of this poor start, the eventual manager was forced to distance himself from the schedulers and work with the project managers. He viewed them in a support and training role. By the time the project was completed, over half of the schedulers had either left the company or transferred.

GUIDELINES

- **Check out a project before joining or taking over.**

  If you are considering volunteering for a project or are a potential draftee, do some scouting about the status of the project. Ask some basic questions about the issues, status, and past events in the project. Then you can answer the important question, "How will I make a difference and contribute to the project?"

- **Play many roles, but not that of a specialist.**

  If a project leader is also a specialist, the team has to consider the leader as having two non-complimentary roles—leader and expert. This can lead to confusion when issues are being addressed. In some cases, the project leader should consider suppressing knowledge of his or her expertise to the team and center attention on leadership.

- **Learn about yourself from the way you manage a project.**

  A project applies stress and presents a variety of situations to you. You then respond. Sit back and review what you do. How are you holding up? Use the scorecard presented earlier.

- **Motivate the team throughout the project.**

  Projects are not for sprinters. They are for marathon runners. Coping with issues over time and dealing with management and organization are constant challenges for the project manager. The key here is to avoid being overwhelmed and to motivate the team throughout the project.

- **Control your administrative time.**

  This reinforces the earlier discussion of duties. Gather the information you require to accomplish these tasks along with a list of things to do. When you are not likely to be disturbed, sit down and dedicate yourself to the work.

- **Re-evaluate the project often.**

  Concentration refers to giving attention to issues, resources, and work. However, unless the project is short, if this is your major activity, you may be tripped up on some underlying problem or issue that you had ignored or not thought about. This means that on a regular basis, you should sit back and think about what is happening in the project overall.

- **Drive the work.**

  Don't just monitor the work—drive the work. This includes the work of all consultants and contractors as well as internal staff. This also reinforces the benefits of a team approach.

- **Be clear on what is wanted.**

  Project managers who are vague in directions will receive vague results.

- **Early in the project, establish how you will work with other project leaders and line managers to share resources.**

STATUS CHECK

- What are the best attributes of project managers in your organization? What are the worse attributes?
- Does your organization have a standard approach for becoming a project manager or remaining as a project manager?
- How are project managers evaluated in your organization? Do the criteria involve motivation of staff, addressing issues, and dealing with crises?

ACTION ITEMS

1. Assess some project leaders around you in terms of the questions and key strategies in this chapter. What common attributes do the project leaders possess? How did they become project managers? Are most of the project leaders hired from outside?

2. What is the process in your organization for becoming a project manager and improving project managers? What training and professional development are provided to the staff or to people who are involved with projects? Is training offered on management as well as on administration and use of tools?

3. If you are not a project manager, consider a project for which you would like to be the manager. What are currently active issues and problems? What could you do about these?

4. Assess the state of the project management process in your organization. Are standardized templates and procedures in use? Are projects with different levels of risk managed differently? Or are differences based on size, cost, or duration?

5. For the project you defined in Chapter 1 and expanded in Chapter 2, assume that you are the project manager. What challenges do you think you will face as the project manager?

6. Evaluate yourself in terms of the following:

   • How much exposure do you give team members in reporting to management?

   • How much time do you spend individually with team members vs. group meetings? Spend more time individually.

   • Do you involve team members in addressing issues? Or, do you present the issue and the recommended action for their feedback? This gets at the heart of the question to what extent the team is involved in decision-making.

   • How do you inform the team of project changes? Do you change the schedule and assignments each time some new item emerges? Or do you implement larger scale changes?

   • Do you know how the team members will react to an issue in advance? How much time do you spend thinking about what the team will think?

CD-ROM ITEMS

03-01   Project Leader Evaluation

Use this to evaluate yourself as a leader.

03-02   Project Manager Evaluation

Use this to evaluate a potential project manager.

03-03   Support for Project Managers

03-04   Project Management Assessment

CHAPTER 4

BUILDING AN
EFFECTIVE PROJECT TEAM

CONTENTS

4

BUILDING AN EFFECTIVE PROJECT TEAM

INTRODUCTION

Traditionally, a project team was formed and kept intact for the duration of the project pending any crisis. Once people were assigned to the team, they stayed on the team. The project team was given necessary resources and was held accountable. This often meant resources were dedicated for an extended time—even during periods of idleness or lack of activity.

This approach has changed over the past few years. Business complexity, downsizing, and mergers have contributed to the fact that few people understand the new technology and have in-depth business knowledge. They will probably not be dedicated to and consumed by a specific project. These factors have also increased pressure on the capable people employed in a company.

The changing nature of projects has impacted the team. People are shared among projects. Many team members also still have to perform their line organization duties. Now more outsiders are involved—consultants, suppliers, partners, and customer firms involved in projects. The projects are more widespread geographically. Also, technology has enabled team members to communicate in a wider variety of ways with faster speed. This chapter addresses the new team environment.

PURPOSE AND SCOPE

The goal here is to examine some of the key questions and issues related to managing a project team. Within this goal, the purpose is to help you assemble and maintain a cost-effective team. The scope includes both large and small projects as well as projects that either are short or extend over a period of years.

APPROACH

Forming a project team and then managing it are often cited as two of the most important parts of project management. Yet, these are often the weakest links in the project and contribute to project problems and even failure. Before specific lessons learned are discussed, let's turn to team management and some of the general concepts successfully employed in the past.

When Should the Team Be Formed?

Some have said the team should be formed as soon as possible. However, you first need to know the requirements and schedule of the project. Prior to management approval, contact a few people to determine their level of interest in the project. Choose people whose work habits and patterns you are comfortable with, who have skills that you think you will need, and who perform tasks well.

Here are examples of what can go wrong.

Example: Banking

In a large banking project, the team was formed early. It consisted of ten people. The project scope and direction were set. Work was started. Within a week it became clear that there was not enough work to keep ten people busy. Rumors started flying about waste. Some team members worked on other assignments. Morale started to sink. The project had to be reconstituted with a smaller number of people. Time and money were lost.

Example: Aerospace

In a case involving aerospace, the plan called for staffing to be built up. The project manager feared that if he did not hire the people according to the plan, the budget of the project would be cut. Instead of preparing a revised staffing plan, he hired the people and the same problem occurred as at the bank.

In general, it may be better to hire slightly late rather than early. If you are slightly late, the pressure to get the work out can give a healthy motivation to the team.

Team formation will continue to change throughout the project. Different needs will arise and requirements will change. Teams today are much different from teams of 20 years ago. You are unlikely to be able to keep a large project team intact. Your team will resemble a play or movie in which the cast changes as the plot progresses. Here are some suggestions regarding timing:

- Identify requirements for a small core of the team that will persist in the project. Get these people on board early in the project.
- Determine requirements for other team members, but add them to the team as you go. Get team members as late as possible to minimize the drain on their time, increase flexibility, and reduce costs.
- Develop the mindset that most team members will be working only part-time on the project doing specific tasks.
- If the requirements of the project are fuzzy, delay forming the team. Wait until the project objectives, requirements, and schedule become clearer.

How Many Should Be on the Team?

Keep the core of the team small—usually no more than two to four people. Why so small a number? First, it is difficult to attract good people to projects, given all of their other commitments. Second, it is easier to manage a smaller number of people.

Other specific reasons for a small core team are the following:

- The small core team is easier to coordinate.
- It is possible to devote more individual attention to the team members.
- The members will feel more accountable since the team size is small.
- The chances of having underused resources are reduced.

Example: Manufacturing

In one manufacturing firm, the team started with three people and kept this number for six months. As the system moved into implementation, more part-time people were acquired for installation and training. The total team at its peak was more than 20 members.

Watch for these disadvantages of a small team so that you can compensate for them:

- Any person who leaves the core team leaves a big gap to fill.

- Small teams can be more difficult to manage if the members do not get along with each other.

- In some organizations, power flows to larger projects with more team members.

- What if you take over a team and it has too many members? After you take over the project, start moving some of the people to a temporary status. When people leave the project, don't rush to fill the slots. Let attrition take hold. Morale might fall, but you can compensate by reassigning the work and getting rid of less critical tasks which can be deferred or eliminated. This might be a good time to review the project structure for excess tasks.

Who Should Be on the Team?

The core of your team should be people who have good general skills, but who have a specific skill area that will be required in many phases of the project. The remainder of the team will consist of part-time and temporary members who enter to perform a specific task or set of tasks and then exit.

An insurance firm had a project in which the project manager was an insurance executive. He felt weak in his knowledge of information systems. He then staffed the team with systems people. However, it later became clear that the team had too many of these and suffered from a lack of people with insurance experience. The project manager had to do double and triple duty by filling in for several team members. The manager had to train the entire team in insurance procedures. The team was not as efficient as it could have been due to lack of diversity.

Here are some questions to ask when choosing people to make up the core of the project:

- Where are the areas of fundamental risk and uncertainty in the project? This is where you want help.
- What are the types of tasks that lend themselves to a "jack-of-all-trades," generalist type of person? You want one person like this who is flexible and can be given a wide range of tasks.

Notice that you did not need to ask what technical or business skills were significant. The skills will become evident over time and they will change. However, if you know in advance that a specific business or technical area will have a major role, then at least indicate this to management. Do not even attempt to get someone committed to the project full-time, since it's not likely this person would be released for such a period. What you will want is to have them work on the project intensely for a specific shorter period.

How Do You Get Team Members?

Make some initial informal contacts to determine availability and desire. The next step is to approach the managers of your candidates. If the people you seek are very good, their managers will be reluctant to let them go. Also, people will be hesitant to leave a secure line position or another project for a more uncertain future in your project.

How do you cope with factors such as these? First, describe to management what makes your project interesting and important. Second, indicate what steps you have taken to ensure that only a reasonable amount of risk exists. Finally, be willing to settle for part of an employee's time. If people join the project team, become interested, and understand that their work is critical, they sometimes become full-time on their own.

You will seldom get all of your first choices for a team. Rather than settle for mediocrity, consider leaving a position unfilled. This offers an opportunity to use volunteers as the project takes off later. Base this decision on how crucial the missing role is at this time.

A project offers the opportunity for project team members to gain exposure with management. This often offers employees a greater career opportunity than they would have in a line organization. This appeals to the self-interest of the team members. Use this approach to attract junior staff to the team.

Temporary team members enter the team to perform a specific set of tasks. When their task is completed, they are either released or they may perform other work on the project. These people can be contract workers or employees. Many times today you deal with contract or consultant people on projects. In one large government project, more than 75 percent of the total team was composed of non-employees. How can this be managed effectively? Employ the old strategy of divide and conquer. That is, manage the work by task area. In the government project, any given area had only one or two contract people, which made it easy to track and manage the work.

Temporary team members must be given an understanding of the beginning and end of their work at the start of the assignment. It is here that milestones must be well defined. Lack of clarity is an invitation for overrunning the project.

Instead of recruiting the top workers, go after more junior people at the start of the project. If you choose team members who are critical to their line organization and other projects, you could cripple their other work, especially if you use them full-time. Also, early in the project you don't know exactly what you need. Thus, it is better to recruit junior workers for a limited time. This gives you greater flexibility and a a chance to evaluate their fit with the project. It allows you to buy time so that you can return and ask for additional people on a part-time basis later.

Training of the Team

For many projects, no training is needed because the people have all worked on previous projects. Even so, consider what training in project management should be given. Each team member needs to understand your methods for resource allocation to tasks, for identifying and resolving issues, and for using computer-based tools in a consistent way.

MANAGING THE TEAM

A key to managing the team is managing the project issues. The work on project issues can be thought of in terms of individual tasks and joint tasks. Joint tasks are those which involve multiple members of the team. Some people favor individual tasks for accountability.

Others recommend that you use team meetings and team effort to do work. An ideal position is to balance the two extremes. Have people spend much of their time on individual work, but also employ group discussion and meetings to address and resolve issues.

Example: Insurance

In a recent large insurance project, each team member had specific tasks to perform, which were monitored. When an important issue arose involving scheduling, technical problems, etc., everyone met for no more than an hour and a half to discuss the issue. Individuals were praised who contributed to issue resolution as well as to milestone achievement.

This approach worked because people understood the importance of the issues and got an opportunity to see why the solution was important and what the implementation steps for the solution would be. People gained a sense of common purpose. After the meeting they went back to their desks and worked on their specific tasks.

When managing a team, get in the habit of holding issue meetings. These are much more important than status meetings or general project meetings. When the issue meeting involves a specific tool or method, use the meeting as a way for more seasoned members to discuss their views and experience.

Get feedback and suggestions from the team. Ask each person what he or she needs and what would be helpful to carry out the work. If an individual provides information, be prepared to act on ideas or problems. Do more than just thank team members for their views. Get back to them with specific actions. Test new ideas. If you use someone's idea, give the person credit.

How should you assign work to people? Some managers assign a few specific tasks to each person—like piecework. They think that this approach will keep a person focused. However, this can lead to boredom. Instead, assign groups of tasks that must be addressed in parallel. At a given time, a person will work on one of these tasks, but he or she will work on all of the assigned tasks over the period of time, such as a week or a month. An example of this method is found in a computer operating system. An operating system works on foreground (high priority) tasks as well as background (lesser priority) tasks. Help employees balance their time between foreground and background tasks by using the issues meetings to clarify which tasks are very important to the project.

In the case of people who bring up personal problems, if possible move them to a flexible work schedule to free them up for a few weeks to address the problems.

How Does a Manager Keep Team Members?

Managers sometimes use bonuses, gifts, and lunches to keep team members motivated. These are fine as onetime fixes but may be ineffective in long-term projects. People tend to expect these as part of the job. Incentive value is lost. Another imperfect method is constant praise. If you keep telling people they are doing a good job, praise will eventually have little impact.

Here are some ideas for managers that will work to keep team members involved for the long haul:

- Involve people in the implementation and resolution of issues. Give the group specific praise for the issue. This shows that you value results and contribution over just hard work. They become more committed if they see that their role is important.

- Minimize the hassles of project management. Help the team members by meeting with their management when needed. Reduce status reporting to a minimum.

- Keep the project team informed of upcoming issues. Give them some insight into the world of project politics. This will capture their interest and give them some idea of what is going on in the bigger world.

- Try to keep a sense of humor.

- Give examples and war stories of past projects to show perspective.

- Keep the administration of the project low key and invisible. If you keep stressing administration, you will lose the team's respect.

- Never compare your project with any other specific current project. However, you can compare your project with other projects generally. Stress why your project is different.

What do you accomplish by doing these things? First, you convey to the team members how much you value their contribution and how much you want to have them involved in the project. Second, you provide them with a view of what you do as a project manager. This will tend to increase understanding. Many team members who have never been project managers mistakenly categorize the job as administration.

How to Solve Specific Team Problems

Observation over the years shows a number of problems that recur again and again in different types of projects. Here is a list of these, with suggestions on what to do when you encounter them.

Problem #1: You have absorbed a team member you do not want.

Upper management may stick you with a "turkey." What do you do? Instead of acting in a way that will show your attitude, look at the problem in a positive way. Determine the person's strengths and assign a noncritical set of tasks. Involve the person in meetings on issues. If this person proves to have valuable skills, continue to assign tasks and increase the responsibility involved in the tasks.

Problem #2: You have to replace someone.

Focus on having team members produce some milestone every two weeks. This will build momentum for the project and morale. If team members attempt to stretch the work out, get into the detail and narrow the scope of their work. Convey a sense that the project is changing and in transition. This is easiest to do with a part-time member of the team. Replacing a full-time member is a major issue. Divide up the member's work among a number of part-time people. This will avoid the team member resenting an individual replacement.

Problem #3: You have an enemy in your camp.

This is the team member who reports what is going on to managers and staff who are hostile to your project. This is very dangerous. How do you counter this? First, work to disseminate correct information to all team members, including those who are hostile. Second, establish direct contact with the line manager to whom the problem employee reports and have regular meetings to go over the project. Third, make the effort to meet with the employee and find out the source of these problem symptoms and get them resolved.

Problem # 4: A team member is not what you thought.

You thought that a certain team member was someone who really knew the technology and systems that were to be used in the project. But it turns out that the team member lacks in-depth experience. If this happens, what can you do? Cover the missing skills. Look for a part-time person who can perform the work. Try to have the team member work with this new person. If this fails, consider moving the team member to other tasks.

Problem #5: Two or more team members don't get along.

This is encountered often. Keeping the team small prevents some of this because there are fewer combinations of human relations. However, it can still happen. You want people to be individually responsible for work, then get together to work on issues. It may be that hostility surfaces at these meetings. Don't gloss over this or ignore it. Take a direct approach. Here is one: "We know that some of you don't get along and we recognize that this is part of human nature. This project is not going to solve problems with interpersonal relations. However, we have to tolerate each other to some degree to get the project completed. So let's make the best of it." In the meetings don't take sides on a personal basis. Keep the focus on the issue. Another action to take is to assign a task jointly between the two members who do not get along.

Problem #6: People become burned out.

Deadlines are tight in many projects today. Resources are limited. The same people are called upon to sacrifice their personal lives and work overtime and after hours. What you can do is take an active role in managing the overtime and extra work. Do not allow it to continue for an extended period. People will start to disappear. Absenteeism will increase. Productivity will plummet. Intersperse periods of heavy activity with forced periods of normal work. Do this even if the schedule has to suffer. As the manager, consider what can be done with the structure of the schedule to make up for the time. Build sympathy for the team with management so that they are aware of the heavy contribution being made. A rule of thumb is that the periods of heavy work should not exceed one or two weeks. Then there should be a two-week period of normal work.

Problem #7: Team members want to work on more interesting, but less important, project work.

If you force team members to work 100% on the important work, they will become resistant and will not work at all. Instead, go to them individually and ask what percentage of time they would like to spend on each activity. For the interesting work assignments, define precise, deliverable milestones that can be measured.

This problem can sometimes be headed off by assigning a range of work at the start or by making weekly assignments.

Problem #8: Work is reassigned.

During any substantial project period, issues arise and changes occur. The project team must be flexible in responding to these new demands. At the start of the project, indicate to the project team that assignments can be changed. The direction of the project may change. Indicate that you will warn people of impending change as much as possible within the bounds of your knowledge. Also, inform team members that you will have fewer, larger changes instead of many small changes. Many small project changes or continuous change can unnerve the project team members and make them feel that the project is adrift.

Problem #9: The fate of the project rests on the shoulders of one employee, who is overwhelmed with critical tasks.

This problem is common and often occurs in cases where only one person has certain critical technical or business knowledge. This occurred, for example, in a natural gas distribution firm where only one person knew how the gas distribution system at a plant was designed and why.

Can this problem be prevented? At the start of the project, ask yourself what critical business and technical knowledge will be needed. Then try to find several people with these skills.

However, it may still happen that one person is critical. What do you do? Sit down with the person and indicate the bind the project is in and that his or her knowledge is critical. Ask the team member what help can be given by others. Ask what else is required to facilitate the job. Your objective here is to have the team member participate in working out a solution, based on the team member's unique knowledge and background. You can assign junior team members to work on tasks with senior team members. This apprentice approach has been successfully used throughout history.

Problem #10: Management wants to change the project team in the middle of the project.

Management wants to remove a key person from your team. How do you respond? First, anticipate that this might happen when you are selecting team members at the start of the project. Assume the worst—that a member will have to leave at a critical time. Plan ahead by having team members do critical work in the early stages of the project, if possible. Second, when the request comes in, don't argue. Instead, develop a constructive transition plan.

Problem #11: Staff productivity is low.

Ask yourself why the staff is not productive. Go beyond the emotional and political areas. Consider what else they are working on in the project. Consider competing projects as well as non-project work. Also, consider whether they have the entire set of skills needed to do the work. They may be trying to learn and to do the work at the same time. Spend time with the staff to find out what is going on. Your last resort is to restructure the work and narrow the tasks that they work on. Identify more near-term milestones.

Problem #12: New skills need to be taught to the team.

Don't feel that you have to train everyone at once. The effect would be to lower productivity overall. Instead, have two people learn the tool or skill. Then have them apply it immediately after they get the training. Set up a meeting in which they give their lessons learned to the rest of the team. If people know that the skill will be used immediately, they will absorb more during training. If they know that they will be discussing it with the team, they will be motivated to master the material and present it clearly. The learning curve for the other team members will then be reduced.

Problem #13: The subteams have difficulties.

In many projects, several people on a team are assigned to work on a specific set of tasks together (forming a subteam). These efforts often get off to a rocky start and have to be redirected later.

Here are some suggestions. First, get the members of the subteam together. Go over their roles in the subteam. What will each person do? Who has overall responsibility? How will they work together? Why were they put together? All of these questions should be asked and answered. Then get the subteam back together when an issue appears involving the subteam. Use these meetings as opportunities to observe and ask how the subteam is working.

Problem #14: Task interdependence delays work.

The result of one person's task is required by another before he can begin his work. This is a recipe for trouble. Head this off in advance, if possible, by trying to eliminate these strict dependencies. If you must have dependencies, ask team members to plan what to do if another member is late. Get together with the dependent team members. Ask what one member can turn over to the other now so that he can start his work. Have them work together to become familiar with what the other is doing. If one encounters an issue, get the other involved in the process of resolution.

EXAMPLES

The Railroad Example

During construction the railroad was a series of projects. People who proved their worth in one project were then redeployed to other projects. Small teams were found in firms that took a narrow view. The project team consisted of the overall project engineer, the logistics manager, the purchasing agents, and detailed site managers for specific lengths of track.

In firms that took a wider view, the project team was larger. Consider the Illinois Central. It viewed itself as a land company that happened to build railroads. The Illinois Central was very successful because the scope of its project was much wider than the railroad's scope. Illinois Central management realized that they had to sell the land and populate it so that there would be demand for the goods to be shipped. Their project team included agents who advertised for settlers and immigrants from Europe and elsewhere. Some lines used agents who received commissions. This would be equivalent to the use of contractors and consultants today.

On the construction site itself, the chief engineer was responsible for overseeing the welfare of the men involved in the work. At any given time this could be the size of an army. In fact, the Central Pacific Railroad used more men than any other project in peacetime in the entire history of the United States. The most successful management examples were those in which workers were treated as individuals, with respect. This humane treatment influenced aspects such as working conditions, sleeping quarters, and food.

Modern Examples

With a multinational project, the manufacturing firm required a project strategy for implementing the network. The project manager identified a management steering committee of six managers, only two of which were from headquarters. Part of their assignment was to work on a strategy for communicating between the busy people involved in the project.

The enemies of the new project were the current professional schedulers who felt threatened by the changes coming. To disarm these critics, two of the junior, more capable schedulers were put on the project team. This had the benefit that potential criticism could be diffused ahead of time. Also, the team had a chance to turn around some of the hostility.

GUIDELINES

- **Look for achievement, rather than experience, when choosing team members.**

    You receive a resume from a candidate with seven years of project experience on five projects. All projects were completed. The candidate looks good on paper, but remember: Project experience does not equate to project wisdom and learning. Find out how the person changed over those five projects. Some people repeat the same errors again and again. Also, the projects may not have had crises, so the candidate existed in a sea of calm.

    It is not the number of projects or the years of experience that are important. What counts is the demonstration of achievement and the ability to deal with issues.

- **Consider apprenticeship.**

    Junior staff are often intimidated by senior staff. Most projects and firms have no apprenticeship program where junior people are assigned to senior staff. The apprenticeship idea does work and should be considered. To handle this, consider sharing of ideas and experiences, as well as apprenticeship. Asking senior people to talk about a particular tool or method is a way for the sharing of experience and lessons learned.

- **Consider asking for only part of someone's time.**

    The people assigned to projects are often those with the fewest current duties. When a line manager is asked to assign someone to your project, he or she might first ask who is available, instead of figuring out who is the best person for the work. Remember, many line managers will get little credit for work on the project. If you ask for only a part of someone's time, you might get a better person than the one most available.

- **Involve as few organizations as possible.**

    For each organization involved in a project, the project leader and team must spend time and energy communicating with the organization. As you add organizations, the burden grows. It grows faster than a straight line since you have cross effects between organizations that you will have to consider.

- **If you inherit the wrong people on your team, make the best of a bad situation.**

  Have you ever wondered, "How did these people get on that project?" You might attribute it to project change or just bad luck. Sometimes line managers put the least experienced and least valued people on the team by intent. If you inherit this, don't spend too much time or energy fighting it. Instead, try to make the most of the situation.

- **Provide an orientation for new team members.**

  It would be ideal for any new person to receive a briefing on the project at the time of joining. This is often not accomplished, however, for many reasons—too much work, deadlines, the person already knew people in the project.

  Orientation can be very beneficial. It can move the person into the right perspective. It can reduce the learning curve. It prevents the new member from plunging in and trying his best with misdirected efforts. The person may then need to be redirected, which wastes time and money.

- **Before choosing a method, think about the skills needed.**

  Any method presumes that the people using the method have certain skills. This applies to basic language skills as well as to complex production systems. When you are considering a method, think about what type of person can successfully use it. If the method requires a star player, and you have few stars, the method is elitist and inappropriate.

- **Assign responsibility, then give team members the latitude of defining how they are going to work.**

  Many project managers direct their team members like line managers directing hourly employees. How much time was put in? What was the hourly output? In most projects this is a portent for disaster. Managing the team by the clock will yield presence, but probably not results. Instead, be flexible.

In one project, a team member worked on the project on weekends. She participated in meetings and worked at a slower pace during the week. This worked out well since it fit her lifestyle. The team was able to accommodate this.

- **Hold one person accountable for a detailed task.**

  Some managers like to assign a task to several people. They write on the project management form or in the software all of the resources involved in the task. The first problem with this is that you can never identify all possible resources required for all tasks. Second, the manager is not differentiating between assignment to the work and responsibility for making sure the work gets done. This act of delegation is very important. Assigning responsibility to one person is best.

- **Clarify team roles.**

  Do not assume that an experienced employee knows what his role is when he is assigned a task. Define the roles of each team member in front of the entire team. This will minimize misunderstandings later.

- **Eliminate excessive project team communications.**

  Excessive communications among small groups can waste time and impact a team's effectiveness in a negative way. Watch for this to occur especially in long projects and in projects in which the team is more isolated.

- **Recognize that the risk and importance of a project lie in more difficult work.**

  Some people prefer to work on easy tasks to build volume. This is human nature. Be aware of this and tolerate it to some extent. The dividing line occurs when team members spend too much time on these small tasks at the expense of the larger tasks. To gain control, ask team members how their critical work is doing. Never ask about the small tasks. This will indicate that the reward structure favors the critical tasks.

- **Avoid polarization of a project team.**

  This can be a by-product of untreated issues. If you leave a critical issue unresolved, it may fester. People individually and collectively share their opinions. The team starts to polarize around the issue into opposing camps.

  Spend time in issue meetings on discussing the issue rather than the solution. If you are correct in analyzing the issue, then the solution usually is more direct. If you don't take any action after several meetings, the team senses a lack of management. If the issue awaits management approval for action, say so. Move on to other issues. Don't beat one issue to death.

- **Avoid giving financial bonuses and rewards for work.**

  This can backfire. For example, a software firm was missing deadlines for development. Financial bonuses were awarded to the key people on the team who were working on the critical tasks. Others saw what was happening and felt that their work was not valued. They slowed down in the hopes of getting a bonus. The project fell apart. The firm collapsed and was acquired by another firm. The software product never made it to the marketplace.

- **Make a place for etiquette on a project team.**

  Many people are sensitive to what is said about them and their work. In a project meeting and within the team, this is especially true, since the team members will be working together constantly. Etiquette and politeness have homes here.

- **Vary project meeting dates and times to increase the level of awareness.**

  Routine weekly or biweekly meetings can lure team members into complacency. Often, the timing of a meeting does not fit with the issues at hand. At other times, a lack of issues encourages team members to revert to small talk. Dump these meetings, as they are generally a waste of time. Consider more frequent meetings when there are many issues and less frequent meetings if things are calm.

- **Do your own project analysis.**

  Some managers delegate project administration and analysis tasks to a team member. They don't want to be bothered by doing such mundane work. This can have several unfortunate side effects. First, the team member becomes more knowledgeable about the project than the manager. Second, the other team members lose respect for the manager. The managers should perform most of the project "what if . . .?" analysis themselves and then involve team members in reviewing and commenting on the results.

- **Take no credit for yourself.**

  A manager who takes all of the credit for the work of the team is eventually a one-person project team. Word will leak out that this is being done. People will mistrust the manager as someone who is trying to get ahead by using the employees in the team. Give credit to everyone. Take credit only for getting issues resolved. Take no credit for the solution or its implementation. Be content with the credit that will come with the results of the project.

- **Review test and evaluation results to raise morale.**

  Most projects have milestones that have to be tested and evaluated. Many project managers lose out on a good opportunity here. They downplay this effort to concentrate on the development or design. But testing and evaluation are very important. They show the team what is passing and what is failing. They also provide a forum for sharing lessons learned. If the test results are negative, in an issue meeting you can go over the reasons for this and how this could be prevented in the future.

- **Focus on progress.**

  Treating staff like children will produce amateurish results. When a manager badgers staff members for status and for work results, he or she is like a teacher who checks students' work every day. What is the alternative? Since people are working toward milestones for their tasks, give attention to what has been done in working toward a milestone. What will come next? This provides indirect pressure on the person to get the work done. Also, you will obtain status by listening to a team member's statements in regard to how the work will be used.

- **Don't ask what you should have done; ask what you should do.**

  Don't waste time looking back except to gain insight for the future. Learn the lesson from the past and then apply it to the future. Don't dwell on the past.

- **Work together to build a common vision.**

  Doing work individually results in the whole being only the sum of the parts or worse. Underlying this point is the conflict between individual and group work. This will be an issue for centuries to come. It is probably a good idea to have a mix of both types of work to develop a common vision of what has to be done as well as to encourage individual initiative.

- **Build a team that long outlives the project.**

  People often gain experience and knowledge from projects. You learn how the business works and how tools and methods are employed. In addition, you can gain friends and form relationships that continue after the project is completed. You can also build on lessons learned.

- **Keep the managers of team members informed of project work.**

  Almost all team members are based in line organizations. When the team members come into your project, they still report and have contact with their line managers. They will likely return to this manager after the project is over. The project manager should treat the line manager as part of the management overseeing the project. This is true even if the manager has no direct tie or interest in the project. You still want to point out the value of this person to the team.

- **Keep in touch with team members who exit the project.**

  Drop by to see team members who worked on a project for months and now are no longer involved. Find out how they are doing. Let them know how the project is going. If you throw a party at the end of the project, make sure that they are invited. Tell management about the credit they deserve for their work.

• **Beware of team members who are very hard workers.**

Everyone tends to think of hard work as positive. It is, if it goes in the right direction. A team member who works too hard and fast can go in the wrong direction quickly and far. They can become burned out.

• **Detect indirect resistance by team members through observation.**

People who do not agree with you often show this through physical appearance and body language. They may look down at the table. They don't look you straight in the eye. They are noncommittal when you ask for their opinions or commitment. They are often silent in meetings. They do not seek you out to discuss problems and issues.

• Consider involving in the project as many people as possible from one department.

Widespread involvement will mean greater support and understanding of the project and its goals. It will also increase support for project results.

• **Work to prevent a project stampede.**

When projects run into trouble, rumors start to fly and people express the desire to leave the project. You may view this as similar to a cattle stampede. How do you turn a stampede? In the western plains, the technique was to move the herd into a circle and then get them to calm down. It is similar in project management. Calm the team members by identifying a series of issues and questions in the project. Focus on these issues and not on the politics. Since most stampedes have a political origin, this often does work. You also may consider more drastic measures, such as replacing one or two team members.

• **When people disagree, depersonalize the situation.**

People who disagree strongly in a project can harbor this hostility throughout the project. The project and the team are hurt.

How do you deal with this? Never allow emotions to get out of hand and personal. Instead, focus on issues. Indicate that many different and acceptable approaches, tools, and methods are acceptable. Also, never announce that one side of an argument is a winner. This will just make the other side angry. Instead, think about how the solution can be presented as a compromise.

- **Mediate in external organizational conflicts.**

  Every team member on a project brings baggage into a project. Team members also bring the position of their home organization into the project. If two groups do not get along and one person from each group is assigned to the project, problems are likely to arise. What should you do? Get the people together and acknowledge the conflicts. Indicate that the project is a different entity. Reinforce this if necessary.

STATUS CHECK

- Does your organization provide any rules, guidelines, or suggestions on roles and duties of team members on projects?
- What is the mix of full-time and part-time team members on projects? Is an effort made to keep the size of the core team small?
- Are lessons learned shared within the team on a project? Are they shared between project teams?

ACTION ITEMS

1. For the project you defined in earlier chapters (or a project selected as an example), identify the team members and their duties. Separate these into core team members and part-time members. Assign responsibility for the detailed tasks of your schedule. What skill areas are you missing? Where do you have gaps? The answers to these questions will provide information on the additional people you will require for the project.

2. If you are currently involved in a project, take this opportunity to assess the project team in the following areas:

Do some members of the team have too much to do while others are not busy? This is a sign that the project team was not thought through or adjusted for workload.

Have you had many part-time members on the team? How are they treated? Is too much time spent getting them on board and later getting them to leave?

4. Sit down with the current schedule and task plan. Compare it with the first approved version of the plan. What are the major differences? How many differences can be attributed to changes generated by team members?

CD-ROM ITEMS

04-01   Team Member Project Evaluation

CHAPTER 5

PROJECT RESOURCES AND FUNDING

CONTENTS

5

PROJECT RESOURCES AND FUNDING

INTRODUCTION

Negotiating for resources and transitioning them in and out from projects are major roles of project management. Resource management ranks in importance right up there with managing issues in a project. In this chapter we will look at some strategies for managing resources effectively.

Resources can be divided into the following categories:

1. **People**—the project leader, the project team, support, and part-time players
2. **Equipment**—machinery, tools, computers, etc.
3. **Facilities**—general and special purpose buildings and rooms, utilities, parking, etc.
4. **Supplies**—major supplies that are not readily available and would be charged to the project.

The last two chapters focused primarily on people—human resources. The extent to which equipment, supplies, and facilities play a role in a project depends on the industry. In manufacturing, equipment, supplies, and test facilities are often scarce and have to be allocated among projects. Construction requires a long list of equipment but less in the way of facilities. In engineering projects, facilities would play a major role. Computers and communications typically are considered equipment, as is software. In the railroad example, equipment for construction and railroads proved crucial, as did supplies.

Equipment and facilities require more management attention than staffing. In many cases, equipment and facilities have to be set up. The equipment may have to be calibrated. Staff may have to be trained to use the equipment. Support may be required during use. When the tasks have been completed, the equipment and facilities have to be moved or reconfigured.

Projects may require many types of resources at different times. Critical resources cost money, affect multiple projects, and require meticulous planning in deployment and use. If you hold on to resources too long, you risk overrunning your project budget. You may also be denying the resources to others. If you are late in receiving resources, you may fall behind schedule. If you fail to obtain suitable resources, the project can be slowed and quality compromised. If you are not politically astute or careful, you can get the wrong resources assigned.

To think of managing all resources across the entire project life cycle is overwhelming. Instead, center your attention on resources in each of the categories that are scarce, significant to critical tasks, or especially importance to you.

Here is a list of questions you can pose when considering a resource. If you answer yes to any of these, you are looking at a resource you will want to manage.

1. Is the resource directly critical to multiple tasks?
2. Do multiple projects require the same resource?
3. Is the resource scarce and difficult to procure or to build?
4. Is the resource complex to use or to apply?
5. Is the resource part of a kit or collection that is critical to certain tasks?
6. If the resource is a person, does the person possess significant skills and knowledge?
7. Are resources being used and shared with non-project work?

Surprises are bound to happen during a project of substantial duration and scope. Resources that are needed through procurement may take longer to receive than anticipated; people are not available when you want them. To prepare for these contingences, go back to the list of issues in the project concept and estimate what the impact would be of the issues not resolved quickly.

PURPOSE AND SCOPE

The purpose of this chapter is to help you reach four goals. You should be able to:

1. Define resources required for a project.

2. Determine the budget for the project.

3. Schedule all aspects of resource management.

4. Acquire the necessary resources.

5. Determine when to release the resources.

The scope of the chapter encompasses all of the resources identified for a project. Activities range from defining needs to acquiring the resources and then deciding which resources to retain. Chapter 12 will cover the actual management of the resources in the performance of the tasks.

APPROACH

Step 1: Determine What Resources Are Needed and When

When the project plan development was covered in Chapter 2, generic resources were listed in the project template. More detailed resources were defined and associated with the detailed tasks to generate the specific schedule for the project. The information in the template will provide the basis for what you are going to do in this chapter.

Consider each of the four types of resources:

- **Human resources**   First determine if the people and skills are available internally in your organization. If they are, you may have to attract their interest and then negotiate for their participation. If you recognize that you require external support due to technical or engineering skills or specific knowledge, then you will be involved in the procurement process.

- **Equipment**   Consider commonly available tools and parts as well as any exotic equipment that may have to be specially constructed for the project. Include software as well.

- **Facilities**   Consider both general facilities, such as office space, and specific facilities, such as test facilities or special storage areas. Included here are utilities, telephone, parking, and other support associated with the facilities.

- **Supplies**   Consider supplies that are chargeable to the project or which are unique to the project.

In all cases in which procurement is involved, be sure to define and place procurement steps in your schedule. The lead time for resources can be 90 to 120 days if you have to generate a Request for Quotation or Request for Proposal, receive and evaluate proposals, and negotiate with the selected winner.

Resource consideration fits in with financial management of the project. Many cases of budget overruns result from keeping resources too long, having resources lie around unused, and mismanaging the resources during the work. In your budget planning, allow for some of these events.

Step 2: Establish the Budget for the Project

Begin with estimating the easy part of the budget. This usually includes facilities, equipment, and supplies. These can be estimated from previous experience. Save for later the more difficult part of the budget, which is to determine the personnel resource requirements. Then carry out the following tasks:

Task #1: Use the project plan to develop a first cut at resource requirements. Get a resource spreadsheet view within the software. The rows are the resources and tasks and the columns are time periods. Export this into a spreadsheet for easier manipulation. Now look at the summary totals for each resource by month. Do these make sense? Are they too low? Often they are. Don't adjust the numbers in the spreadsheet. Instead, go back to the project plan and project management software and modify your resource loading on the tasks. You may even encounter missing tasks. Continue doing this until you are satisfied that you have the major resources.

Task #2: Now take the spreadsheet from Task #1 and add the resources for facilities, supplies, and equipment by task area. This will allow you to determine when you will be needing these resources. Another approach is to include these as resources in your project plan.

Task #3: You are now reasonably close, but you probably want to add some slack or padding to the budget for safety. Do this in the spreadsheet. If you do it in the plan, there is a problem in that the schedule will probably be too unrealistic.

Even though you may be tempted simply to create a spreadsheet and put the budget items in, avoid this because the plan does not match up to the budget overall. Moreover, the budget will not match up to the requirements of when money is needed. If you link it to the plan, you have not only obtained a more credible budget, but also a more credible plan.

Budget for large and multiple projects using a bottom up approach. That is, start with subprojects or individual projects, then aggregate these to get an overall picture.

Step 3: Create Project Oversight

Unless your project is small in budget, resources, and time, you should have management oversight for the project in an organized way. This will provide a basis for dealing with issues and opportunities as well as a communications mechanism for management relations. It is best to consider a small steering committee. This will be easy to create.

Many managers are already overworked and overcommitted. It will be difficult to attract good managers. Look at the project and determine which departments are going to be involved. Go to the departments manager with high level tasks and milestones identified. Indicate the reasons they should be interested in the project. Point out that the committee will meet only for major milestone review and for issues that could not be solved at lower levels. This will show that you respect the various demands on the employees' time.

After getting some interest, hold your first informal steering committee meeting. Give the members of the committee an overview of the project and budget. Present the list of issues from the project concept. Do not ask for any decisions. The initial meeting should take no more than one hour.

Step 4: Integrate the Resources in the Project Schedule

When you have identified the resources needed, schedule the following tasks in your plan for each resource:

Task #1: Determine and document specific resource requirements.

Task #2: For internal resources acquisition—

- Identify resource candidates and determine their avail
ability.
- Negotiate for internal resources.

For external resource acquisition—

- Prepare necessary requests for external procurement in terms of schedule, duration, and requirements.
- Procure external resources.
- Negotiate for external resources.

Task #3: Prepare resources for use in the project after acquisition.

Task #4: Determine the release date for each resource.

Task #5: Prepare the resources for release.

Task #6: Release the resources.

Task #7: Follow up on open items after resource use.

List these tasks in your schedule for each type of resource, even those that are not readily available, to avoid missing tasks as the project progresses.

Adding all of these tasks to the schedule serves a political purpose as well as an organizational one. Management will be aware of the effort required so that they can provide support if the procurement hits a snag.

Step 5: Define Your Resource Strategy

This may sound like a vague step. However, most of the problems encountered in getting resources can be traced to a lack of thought and consideration early in the project. Don't assume that just because the budget was approved resources will automatically be assigned. These are two different steps involving decisions.

What should be your resource strategy? The first part of a strategy is to concentrate on resources required over the next three-month period. This will get you started. Don't ask for resources further in the future. Commitments may be meaningless since you haven't yet shown results.

During the first few months, the project should begin to show results as initial milestones are achieved. Momentum will build. With this progress you can move to the next part of the strategy. With results in hand, approach management after the first month to begin to seek approval for resources in the next four to six months. These time frames are flexible, depending on the specific project. Continue with this pattern. You are seeking a rolling commitment based on continued results. This is much easier to accomplish than wholesale commitment at the start of the project. Management will also feel more comfortable because they will have greater control.

At this point, establish priorities for trade-offs. Will you take a less desirable resource as a trade-off for lower cost? Are you willing not to have the resource when you want it in order to get the desired vendor? Get your priorities straight.

Prepare for positioning the project in terms of the techniques needed in negotiation and whether you get access to the best resources. Do this by answering the following questions:

- **Who is responsible for the resources that you require?**

  Find out both who has direct responsibility and who has political responsibility.

  For internal resources, begin contacting managers who control the resources at the start of the project or very early in the project. If you fail to do this early on, you will upset the plans of these managers when you require the resources, you will upset the plans of these managers.

  For external resources, contact purchasing to determine the process for acquiring the resources, the various steps, and the schedule. Start building rapport with the staff in purchasing.

- **What other projects and work are demanding the same resources?**

  Remember that your project will be going on for some time. If your plan calls for a resource from the second to the fifth month of the project, consider any project that requires the same resources from now until the end of the sixth month or longer. Allow for slippage.

What are the benefits of the project to management and their organizations after the project is completed? What is their self-interest in giving the resources to you? What are their objectives and goals?

You may already know people in upper levels of management. If possible, sound them out on the potential problems of getting the resources. Maybe they can introduce you to the managers of the needed resources, so you do not have to make a cold call.

Step 6: Win the Competition for Resources and Complete the Procurement

With downsizing and rightsizing, many organizations do not have spare people, equipment, or facilities lying around. You may have to compete with other projects for resources. If you are the manager of the key project in the company, then you get priority. However, this is clearly the exception. The general situation is that you head up one project among many and the world will not end if the project is not completed. How do you compete?

A first guideline is to make sure that you have developed a realistic minimum requirement for resources. Tell managers that this estimate is truly a minimum and is realistic. You will have to be willing to negotiate for resources. Be ready to trade off.

A second guideline is to employ the project team as part of a sales force to obtain the resources. This approach is preferred to appealing to upper management to force a line manager to release equipment or people to you. That approach will breed hostility because you have removed the resources from the control of the line manager.

A third guideline is to follow your resource strategy. Aim at incremental commitment. You will tend to attract more support and resources with the momentum of success.

Let's turn to procurement. Provide purchasing or the line manager with the resource requirements, a copy of the project plan and schedule, the specific tasks that you desire to be performed, and the milestones or end products that you seek from the resources. Indicate how the tasks affect the overall schedule. The more information you provide, the more comfortable the managers and staff will be in trying to help you. The more vague you are and the less information you provide, the less likely timely cooperation and support will be. Nail down your agreement in a memo of understanding that can help resolve any problems later.

During procurement, be available to help. The following tasks will fall upon your shoulders. Respond quickly.

- **Preparation of the statement of requirements or statement of work to be included in the Request for Proposal.**

  This statement should specify what goods and services are to be provided, how they will be used, the end products expected, the schedule, and how the resources will be managed.

- **Participation in answering questions at bidder's conferences.**

  The purchasing agent has no detailed knowledge of what you are trying to do. He or she will pass questions and inquiries back to you for response. Timely, accurate, and complete responses are called for.

- **Review of proposals to do the work or provide the goods.**

  Provide resources for this review. It may involve you having to search out staff to review these with you. When you select the winning proposal, always line up several backup vendors in case the later negotiations do not succeed.

- **Support in vendor negotiations, if needed.**

  In most cases, the purchasing agent does not want you involved. However, be on standby in case there are problems or additional questions.

Step 7: Plan for the Transition of Resources into the Project

If you are bringing people on board, establish some kind of orientation for the project. Don't expect people to jump in without some sense of priorities, tasks, and the project plan. Never assume that they were prepared or briefed by their own management in advance. Cover basic issues such as where they will work, how they will access the building, what telephones are available, and how parking will be handled.

If you are moving equipment in, make sure the necessary support (utilities, space, support staff, etc.) is in place. Many times equipment arrives on time just to sit on a loading dock for several weeks because the organization was not prepared to receive it. The transition also may include personnel being trained in the use of the equipment.

Facilities can present challenges. Visit to see if the facilities are ready for your use. If additional work is to be performed to get the facilities ready, who will manage the work? Who will do the work? Who will pay for the repairs or cleanup?

Step 8: Transition the Resources into the Project

This is the actual transition into the project. It is possible that you may find problems immediately with the resources. The right people were not sent. The equipment does not work. The facilities have power problems. If this should occur, contact purchasing, management, or the vendor. Stop using the resource. Do not use the person on the project. Do not employ the equipment. Do not move into the facility. Usage can be interpreted as acceptance.

Step 9: Determine When to Release the Resources

Even as the resources are beginning to be employed in the project, define how and when resource usage will stop. Make sure that this is in your schedule. Develop a turnover approach for the release of the resources. In most cases this can be a simple checklist. Never allow the core project team members to accept these temporary resources as permanent. Also, reinforce the temporary nature of the work with the people who are brought into the project so that there will be no misunderstandings.

The release of resources should be announced to the project team. For people, you will want to debrief them and get their lessons learned. For all resources, obtain feedback from the project team on how the resources could have been put to better use and the lessons learned.

In your budget analysis, consider what percentage of the time the resources were effectively used. This can help pin down any budget variances. It can also bring home to the project team the cost for resource waste.

If you have an opportunity to release a resource without significantly harming the project, then do it. Remember, the fewer resources you have in general, the better.

HOW TO COPE WITH CHANGING REQUIREMENTS

Here are some reasons that the resource requirements on a particular project may change after the project begins.

- You find new information as you progress in the project.

  This may alter the resources needed. In the railroad construction example, if the construction crew runs into granite rock, the train track may have to be rerouted or the crew may have to obtain additional blasting materials. If this discovery resulted in the crew members falling behind schedule, they would have to obtain additional workers to compensate.

- Management shifts requirements or direction.

  Management decides on new requirements that the project must address. This can change the nature of the project entirely, including which resources are suitable. Changing requirements may occur in software systems, engineering, or marketing projects, for example.

- A change in resources results from external information.

  A competitor is about to introduce a product better than the one you are building in the project. Your team is sent back to the drawing board for a new product. This may mean new resource requirements. Another external source of change is government regulation. A rule can change, impacting the underlying assumptions of a project. The same is true with new technology, which may replace technology currently in the project. This change may call for different staffing and skills.

- You have alternations in timing of requirements and the amount or extent of the resource needed.

  The digging of the Panama Canal is a classic example. Changes occurred because equipment and facilities needed were initially underestimated.

HOW TO TAKE ADVANTAGE OF RESOURCE OPPORTUNITIES

An opportunity in this context occurs when a resource that is potentially useful to the project suddenly becomes available. To take advantage of this situation, do the following:

- Keep an eye out for resources at all times. Alert managers and your project team that you are always looking for resources—"A few good people."
- When you hear about potential availability of a resource, analyze your plan to see what benefits you could reap from this. Also, assess the financial impact on the project.
- Be ready to sit down and negotiate for the resources immediately. Cut a deal.
- Put the additional resources to work immediately. You will not only appear organized, but you will also attract more opportunities later.

EXAMPLES

The Railroad Example

The late nineteenth century was a period of unprecedented railroad expansion. Some specialized jobs were hard to fill. Equipment had to be hand-built specifically for railroad construction. Yards and other facilities were necessary for maintenance and work on locomotives and rolling stock. Most railroads managed to deal with these resources well. How did they do it?

Railroads were the first to put in place procurement procedures for remote and local managers so that the managers were given certain autonomy to address problems within their area. In addition, the railroad itself was employed to shift resources rapidly from one location to another.

To address manpower requirements, the railroads used not only recruiting campaigns but also apprenticeship programs. Some railroads pushed their employees to learn more and move into higher paying, more skilled positions. Several railroads began informal societies involving employee spouses as a means of raising morale and also for improved welfare. Specialized skills and knowledge were passed on through job rotation and transfer so that the body of knowledgeable employees was constantly growing.

To generate equipment and ensure adequate facilities, many railroads started other companies that made equipment (foundries) and constructed facilities and railroad yards ahead of the time of their use.

Modern Examples

At the manufacturing firm, people in remote locations began to charge their time to the project without the leader's knowledge. The project leader could not be in many different countries at once. A subproject leader was appointed in each geographic location to handle administration and coordination. This shows that even with the Internet and rapid communications, it is important to have local control.

The project manager at the consumer products company found that high priority projects did not receive resources when needed. However, lesser priority projects did have the resources. What was wrong? No method was in place for assessing the resources on a project at any given time, so it was difficult to redirect resources between projects. Large, priority projects had project leaders who considered resources individually and did not work together.

GUIDELINES

- **Identify at the start as many resources as might be needed.**

  The more comprehensive your list, the lesser the chance you will be surprised later with a new requirement. Review your resource list once a month and update it.

- **Develop a transition strategy.**

  Not only must you have a good understanding of the resources required, but also you must be able to transition resources skillfully and take advantage of opportunities. You must be able to come up with creative substitution approaches.

- **Seek incremental commitment from management.**

  Don't request too much over too long a period. Also, keep management up-to-date on the project and maintain interest in the project.

- **Make sure that your approach to resources is integrated.**

  Consider resources in groups necessary to perform and support specific tasks. This is an integrated approach. If your approach is on the basis of individual resources, you will be more likely to miss some resources. You will also lack focus during negotiations.

- **Think, plan, and take small actions when a resource is added.**

  Often this is wiser than precipitous, decisive action. Employ the new resource on a trial basis to see the results. If the resource is a person, for example, and you take decisive action and hand over a task area to the new resource, the person may fumble around for weeks and hurt the project.

- **Do not treat all projects equally.**

  Equality produces mediocre results. In considering resource assignment the first thing to discard is fairness. Projects have different levels of importance and benefits. No two are alike. Therefore, it doesn't make sense to treat them the same.

- **Keep infrastructures small.**

  Every project has an infrastructure. Included are files, methods, policies, procedures, and tools that are used in the project. In addition, projects have a project manager, a project organization, and project support. It follows that many times more infrastructure means more control and structure. More structure may mean that decisions and action in the project take longer. Smaller infrastructures are more efficient. The Romans were able to build large aqueducts in months with a small, organized team. The aqueducts were well built, many lasting for more than 1500 years.

- **Take on and manage some unsuitable people to contribute to a project's political success.**

  In an ideal world, you could assume that only the best and brightest will participate in your project. In real life, you run into mediocrity. Also, your project may be the place a manager unloads an unwanted employee. Your first reaction might want to fight it, but at what cost to the project and your career? Keeping that team member may be politically beneficial. Look over your project and see if you have any slack where this person could do some useful work.

- **Structure the work so that you will not need the best people.**

  Large projects stall under internal competition for resources. Even large organizations may have only a few highly experienced and qualified people. In your plan never assume that you will get the best.

- **Work with less resources.**

  Can you do the work with less in terms of resources? This is one of the questions you should always ask during the project. Ask it yourself before someone asks you. In an era of downsizing and efficiency, making do with less is essential. Remember, fewer resources means less to manage. It is also true that smaller projects get more done than large projects due to less need for coordination and a simpler chain of command.

- **Do not try to hold on to resources.**

  Do not hold on to resources that are not being used. You will become known as a person who hoards and wastes resources. Later, when you make new requests, you will be more likely to be turned down.

- **Understand why people are motivated to be on a project.**

  It is important to understand why someone is on a project. Obviously, many possibilities exist—attention, risk, the desire to do something different, the desire to learn, etc. Even if a person is assigned to the project and has no choice, you can still probe for what that team member would like to get out of it. What do you do with information? You use it to your project's advantage. Structure the project so that the work gets done and people get some of what they want. It pays off.

STATUS CHECK

- How long do people stay on projects in your organization?
- Are efforts made to release resources as soon as possible?
- Do your projects have resource strategies?
- Are efforts made to get management to commit resources too early? Are many resource changes needed? Is there often a mismatch between the resources you have in the project and those that are needed?

ACTION ITEMS

1. For the project you developed in the first chapters, make a more complete list of people, equipment, and facilities that you require on the project. For each category of resource, identify the manager or vendor who can provide the resource. Identify alternatives to be used if the selected resource is not available.

2. Using the information from the list in Question 1, develop a GANTT chart in which the tasks are resources arranged or sorted by type. The schedule is the time that the resource is needed. This GANTT chart is a useful tool to help you plan resource transitions.

3. Return to your project plan and insert the resource-related tasks that were identified earlier in this chapter.

CD-ROM ITEMS

05-01   The Project Budget

05-02   Sample Common Resource Pool

CHAPTER 6

METHODS AND TOOLS—
THE NUTS AND BOLTS OF HOW

CONTENTS

6

METHODS AND TOOLS—
THE NUTS AND BOLTS OF HOW

INTRODUCTION

A method is a technique applicable to project management. A tool supports the use and implementation of the method. In the past, few choices in methods and tools were available. These included the critical path method, the general method of creating a schedule, and methods for analyzing a project. Each project was a one-person show. Prior to software, the tools were paper forms. Information updating was a labor-intensive process consuming massive clerical labor and white-out.

The arrival of PC's and project management software changed the tool landscape, but did nothing for methods. Even though maintaining schedules was less tedious, the overall methods still focused on a single scheduler or manager doing the work.

Three generations of project management software can be distinguished:

- **First generation.** These were tools based on mainframe computers. The scheduler obtained updates to schedules based on a batch-processed turnaround report that the scheduler distributed. The scheduler collected the updates and input the data into the computer. The computer produced batch reports that the scheduler then analyzed. Errors were corrected through additional input. With the "final" reports in hand, the scheduler then distributed the reports along with a new update sheet.

  Problems with this process became apparent. First, by the time managers received the information, it was out-of-date and virtually worthless. Second, managers had to create their own schedules to address their own analytical needs. Nevertheless, impressive graphs and reports were produced. You could walk into a planning or conference room and see a PERT chart that covered all of the walls in the entire room. Thousands of tasks were printed courier font of size 10—virtually unreadable, but impressive.

- **Second generation.** With the arrival of PC's, early, crude software became available. These software tools allowed limited drawing capabilities, little analysis, and were very restricted in terms of the number of tasks and resources. Individual managers could produce simple GANTT and PERT charts. Since the mainframe project management systems continued to be run, the end result was often confusion between the two systems due to incompatible information. Those of you who have been around for awhile may remember VisiSchedule and MacProject.

- **Third generation.** With advances in PC's, people finally were given industrial-strength PC project management software. Microsoft Project and Symantec Timeline were two of more than 20 such packages. This generation of software tools led to the demise of the mainframe project management systems for many organizations.

  However, while the tool changed, the method often did not change. There was still the scheduler who worked now at a PC. While this was an improvement, the underlying problems persisted. Project managers continued to manage their schedules. Multiple schedules abounded.

  The requirements of project management are for methods and tools that support sharing of effort and collaboration in scheduling. While tools changed, their new benefits were largely mitigated by lack of change in methods. Organizations failed to have an effective strategy to allow project management to use new methods to take full advantage of the newly available tools.

- **Fourth generation.** People are now finally starting to see the emergence of collaborative software tools. These are software products that extend project management across a network and allow updating, scheduling, and issues to be addressed in the network with aspects of groupware and electronic mail.

  Still missing are the methods to use such tools. That is going to be a major focus of this chapter. What will happen if firms adopt the software, but not new methods? They will fail to realize the benefits of the tools. Without new methods, things will probably be worse than before the new software was adopted.

PURPOSE AND SCOPE

The purpose of this chapter is threefold. We will:

1. Define a method and tool strategy;
2. Define new methods for project management that take advantage of modern and emerging tools; and
3. Provide guidelines and tips on how to use the methods and tools.

The purpose of this chapter is *not* to tell you how to use a one-person project management system. This information is available in dozens of books and manuals.

The scope of the effort extends beyond project management software. It also includes the Internet, Web, intranets, groupware, database management, electronic mail, and other software. The goal is to help you be successful through more effective and innovative use of modern methods in conjunction with network and software tools.

APPROACH

A method is a technique, process, or procedure that supports project management.

A tool is something that supports the implementation and use of the method. Typically, this means software tools. Other tools include presentation aids and facilities. Manual filing systems are another category of tools.

Tools shape methods and vice versa.

BENEFITS AND PITFALLS OF METHODS AND TOOLS

A method provides a standard way of doing something. This is particularly important in organizations in which people undertake multiple projects with different people. If everyone does something different, there will be chaos. This can also happen with tools, although with less impact.

Beyond the benefits of standardization, common tools and methods can provide a company with a rich source of information from projects and from lessons learned. Incompatibilities, on the other hand, reduce this capability.

A third benefit of standard methods and tools is predictability. People know how to behave and how to use the methods and tools without additional training. They can come up to speed faster for the new projects.

Watch for these potential pitfalls with the use of methods and tools:

- If the methods and tools are not synchronized, benefits are reduced.

- People tend to be dazzled by a new tool. They adopt it without thinking of the method. Everyone adopts it a different way. Chaos reigns.

- People can become too comfortable with a set of methods and tools. Without thinking, they apply the methods and tools to projects of all sizes and shapes. The result is overkill and small projects become swamped with overhead and process.

- People resist new methods and tools.

Example: Retailing

A retail firm had 12 different projects going on simultaneously. The firm was trying to implement EDI, store point-of-sale, conduct video training, perform scanning, and handle management controls, all at the same time. While all the project leaders followed the same method, they used a total of four different project management software tools. They also had set their schedules and plans up differently so that it was impossible to get an overall picture of what was going on. One project failed and dragged down two other projects. It was like dominoes. When the firm attempted to standardize its tools, it was too late.

A WINNING STRATEGY FOR USING METHODS AND TOOLS

You need a strategy that achieves the following goals:

- The first goal is *scalability*. That is, you want to be able to apply the method or tool to small as well as large projects. Also, plan for use with single or multiple projects.

- The second goal is *collaboration*. You want the method or tool to be compatible with having multiple people work with it in the same or different projects at the same time.

- *Modernization* is a third goal. You want a strategy that will accommodate new methods and tools in a smooth manner.

- *Measurability* is another goal. You want to be able to measure the effectiveness of the method or tool so that you can be assured that you are employing it effectively.

- *Formulating lessons learned* and improving at project management over time are also major objectives.

A practical general strategy is to have a standard set of methods and tools that address the goals above for all projects. Beyond this, supplement with additional methods and tools for specific classes of projects (e.g., very large projects, projects with specific customers). The strategy would also include processes for capturing and using lessons learned and experience. A regular measurement assessment process can improve the strategy.

WHERE AND WHEN METHODS ARE NEEDED

Here is a list of method areas to address in the plan and suggestions for each area:

- **Setting up the schedule and plan initially**

  The method should include a project plan template (see Chapter 2), along with lists of tasks and resources. Guidelines in the form of a step-by-step process should be given. The review method for the first version of the schedule should be identified.

- **Analyzing and improving a schedule**

  Support this with guidelines and examples. In addition, the availability of an "expert" helps. Look at the wording of tasks, split up long or compound tasks, reduce dependencies, and perform a specific "what-if" analysis.

- **Updating and maintaining a plan**

  Support this with a formal method in which the plan and the detailed project plan and schedule are maintained on a network at all times so that they are visible. Update plans at least twice per week. Label detailed tasks as either not started, in process, or completed. Summary tasks that are a rollup of several detailed tasks will have a percentage complete based on the detail underneath.

- **Modifying and changing a plan**

  Changing a plan in a significant way in terms of schedule, resources, and deliverable items should require management approval. Changing task structure and assignments without impacting the schedule in a significant way does not require such approval. Support this policy with a detailed set of procedures on modification and updating.

- **Addressing issues and communicating within the project team**

  Draw up a common and standard set of procedures for handling and tracking issues. These can be supported by guidelines on how to identify issues. Also provide lists of issues.

- **Communicating and working with people and organizations outside of the team**

  Communications is often a matter of style. Many different project managers have many different styles. Without intruding on personalities, provide information on what organizations outside of the project can expect from the project team.

- **Assessing benefits, costs, and risks associated with a project**

  Have a standard set of rules for defining benefits and costs for each project. Don't deal with fuzzy benefits. Deal only in tangible benefits. The same philosophy applies to costs. How costs are applied to a project is unique to the specific project as well as to the organization.

How do you select methods? Consider what other companies have done. Examine large and complex projects that were successful. This will give you ideas about methods. You should also be able to expand on the methods presented in these chapters. Here are more specific guidelines:

- **Consider a statement of method that uses everyday language.**

   A method that employs arcane words is too much work and effort without the additional payoff. People will not use the terminology anyway since it is not familiar.

- **Pick all methods at the same time.**

   For each of the method areas listed earlier, identify a specific method. Then sit down and determine how someone would work with the total set of methods. Or, do they overlap so that their use becomes counterproductive?

- **Evaluate scalability and flexibility.**

   Here is a test. Choose three projects of different sizes. These can be in the recent past or present. Next, choose several projects that are very different from each other. Apply the method to all of these. How do they scale up?

EFFECTIVE PROJECT MANAGEMENT TOOLS

The fourth (most current) generation of project management tools provides some or all of the following:

- Standard project management features
- The capability of sharing of project information, a plan, and a schedule in a network
- A means to delegate tasks in the project to staff

- A way for staff to update management and communicate with management regarding their tasks

- Consolidation and tracking of all work

Such a software package combines a database of project information, project management software, electronic mail, and groupware. It offers the benefits of an integrated approach for tools. It is simpler than having a grab bag of tools.

First, establish the team. Second, establish the plan on the network with the tool.

If such a tool is not available, assemble a unified set of methods that draws upon specific tools. Integration is provided through the methods and not the tools. Take every item in the plan and define the procedures and tools. You could then expand on this with specific guidelines for small, medium, and large single projects, as well as for multiple projects.

The manager and team members can now work together with methods for identifying and resolving issues. They can track progress and identify areas of risk.

What tools would you employ if the integrated tool were not available? Here is a list of tools and the procedures they define in carrying out project management tasks.

- **Project management software can have data resident on a network to allow sharing of the files**

  Establish the basic project plan on the network with the baseline plan (agreed upon plan), the actual results, and an estimated plan for the future. They can add more detail to their part of the plan. Encourage people to review and update their schedules once a week and maintain their own tasks. The danger is that people make mistakes, so save the file frequently during the week.

  The project plan has fields for the person responsible, whether the task is management critical or has risk, and the issues that pertain to the task. These are put in individual text fields associated with a task. These fields are then searchable.

- **Electronic mail software supports attaching project files**

  Electronic mail can also be used as a vehicle for sharing and routing information. The limitation of electronic mail is that it lacks the database elements of groupware. It is freeform, which will cause problems later when you want to do analysis. It is often better to use the electronic mail within the groupware to add structure.

- **Groupware or a shared database tracks issues and action items**

Groupware allows you to carry out a dialogue for managing a specific issue or updating the plan. Groupware has been employed for several international projects. It was found that the elapsed time to obtain decisions and resolve issues is cut by more than 50 percent. We will discuss groupware in more detail in Chapter 7.

The issues database contains all of the information pertaining to an issue or opportunity. The database is stored on the network so that people can update the issue information. Data related to status, assignment, resolution, description, category, and tasks that relate to the issue are all in the database. You can now put the issues and project plan together. You can search the project plan, filtering the plan to obtain only those tasks that relate to an issue or are dependent on other tasks. Alternatively, you can search the database for all issues pertaining to a specific task. Establish a similar database for both lessons learned and action items. If you employ standardized templates of high level tasks, the results are even better, since anyone who employs the template can access the lessons learned for specific tasks.

- **A spreadsheet can be used for project analysis**

Spreadsheets can help overcome some of the limitations of project management software. Most project management tools allow you to assign costs through resources. You can input standard and overtime rates and vary the calendar for specific resources. Some project management packages support earned value work and financial calculations. However, this is often too inflexible. Establish a table of data from the project database and then export the table to a spreadsheet. The spreadsheet can then be manipulated and analyzed to obtain cost and resource reports.

- **Internet links allow access for remote managers, customers, and suppliers**

  Using the Internet is considered in Chapter 11.

Methods and tools can cover about 60 to 70 percent of what can arise in a project. The remaining percentage, based on human behavior and communications, should not be ignored. A major benefit of the methods and tools lies in freeing up time to address these other things. In project management you need time for establishing rapport with people, negotiating, fact finding, and making decisions. Methods and tools can make the project run more smoothly but they seldom reduce overall time and resources. There is a trade-off between the learning curve and the efficiency gained from the tool or method.

What does the future hold? It is evident that the tools are becoming more supportive of collaborative work. This bodes well for project management. It will eliminate some of the concern that people have for using project management. As it becomes easier to do, you will be able to use the tools for smaller projects. The use of the Internet provides standards for data sharing between firms and supports compatibility. Thus, you should see more effective projects between companies. In fact, it is not unlikely that more projects will be generated due to the availability of tools. The availability of collaborative tools will also mean good project managers can take on more work if their time is freed up. This should encourage more parallel effort.

There is a downside to using methods and tools. If you lack good methods and discipline, these tools can lead to more disorganization. Mistakes can multiply, since communications are faster and the results of work in a plan or database are instantly capable of being viewed by many people. It is a case of the good getting better and the worse deteriorating.

Tool Selection

Begin with the common PC tools that are in place in your organization. The first principle is to eliminate tool candidates that are incompatible. Next, try to use what the company has already purchased. You don't want to fund a major infrastructure software and hardware project out of your project budget. Since there is typically an office suite, electronic mail, database, and a groupware set of approved software, your choice is often limited as to which project management software you will use.

MEASURING THE EFFECTIVENESS OF CURRENT METHODS AND TOOLS

Before you replace or renew methods and tools, assess what you are have been using. Are the current methods effective? Are the tools you are currently using effective? What are the major problems you are having today? Can you measure what results are being obtained?

Here is a list of what you might look for in assessing your current situation:

- How many different schedules exist? Are they being tracked for the same work? If there are many, then a common effort will reduce redundancy.
- Are many project meetings consumed in getting exact data on the status of a project and resolving discrepancies between different plans? If so, then issues are not being addressed.
- How long does it take for an issue to be resolved after it has been identified? If the answer is "Too long," time is probably being wasted while people spin their wheels waiting for decisions.
- What comments do people make concerning the current process? Solicit and collect comments from team members.

Prepare an evaluation report of current methods and tools and circulate this to lay the groundwork for introducing change. Here is an outline for such a report:

I. Introduction—purpose and scope.

II. Overview of current project management process, methods, and tools, describing what is being done today.

III. Issues with the current methods and tools. List the issues and describe their impact on the organization and projects. Use specific anonymous comments from staff.

IV. Summary—the cumulative impact of all of the problems.

IMPLEMENTING NEW METHODS AND TOOLS

As preparation for initiating the new methods and tools, answer these questions:

- How will exceptions to the general methods be addressed? Will people improvise, or are additional procedures necessary?
- Will the methods and tools be applied to all projects? If so, how will they be retrofitted onto existing projects? Who will pay for the learning and familiarity time? How will they be applied to very small and very large projects?
- Who will serve as the tool expert? Will this person be involved in training others? Is the expert going to be easy to reach when help is needed?
- How will the methods and tools used be monitored to ensure proper use? What constitutes proper use?

Here are three steps to use as you implement methods and tools:

Step 1: Start small. Begin on with several small projects. Learn from these projects to build lessons learned and guidelines. Document and use these guidelines and step up to new projects.

Step 2: Build a project with multiple parts and do a cross-project analysis. This demonstrates the added benefits of standardized methods.

Step 3: Analyze and perfect the methods and tools. This way, many more people will be able to use the revised methods and tools successfully.

Here are a few words of caution:

- Keep management fanfare and endorsements to a minimum to avoid bad feelings.
- Do not implement the approach on existing large-scale projects that are far along. This will slow progress as people attempt to use the new tools and methods.

- Test methods and tools so that you have experience with them before asking others to implement them.

The Cost of Implementation

Implementing modern collaborative scheduling is neither inexpensive nor easy, possibly involving hundreds of people and requiring a tremendous time commitment. Using a multi-phase implementation approach will help by spreading the cost over a longer period.

Some of the major cost elements are as follows:

- Hardware, software, and network upgrades
- Application software licenses
- Training and documentation costs for the new tools
- Staff time in learning the new methods
- Management time to measure and evaluate the new tools and methods
- Potential customized software development to provide additional features
- Conversion effort to move current schedules and data to the new tools

Once you implement the methods and tools and have used them for a time, you are ready to move into another part of the evaluation. Since tools support methods, focus on how people are using the methods. Also examine evidence that the tools are being used. Have team members discuss how they are using methods and tools.

Here is a list of questions you can ask:

- Is there consistency between projects? This refers to structure of the plans and whether they follow similar templates.
- Can you assemble an overall schedule composed of the detailed schedules? Will it make sense?
- Have the project managers identified issues and are these being addressed?

- Is there a table of issues vs. projects to see what different projects have common issues?
- Are people asking for help in using the methods and tools?

POTENTIAL PROBLEMS AND OPPORTUNITIES

Even with analysis, organized planning, and necessary support, problems arise with the implementation of new methods and tools. Here are some common problems encountered, along with suggested remedies.

- **Resistance to new methods and tools.** This is natural since some team members may have used the previous methods and tools for years. Concentrate on winning over the younger staff members who are less resistant. As project leader, be a strong advocate of the new methods and tools through the processes of tracking the project, resolving issues, and maintaining communications.

- **Complaints that the new methods or tools take too much time.** Review how the team members are using the methods and tools they are complaining about. Hold a meeting specifically to address how to use these. Have team members demonstrate how these would be helpful.

- **Resistance to formal methods and tools.** Perhaps you are asking for more detailed methods and tools than the team members are accustomed to using. To counteract this attitude, show the team members that the more formal methods and tools free up time that can be better spent dealing with issues.

- **Team members raise new ideas on how to do project management.** Don't discard these. Try to get them reviewed and, if feasible, adopt them. This will show that you care about your team members and their ideas. And, of course, exceptional ideas can come from any team member.

- **A new release of software appears.** Team members are enthused and want to use it. Always be ready to carry out a pilot effort to evaluate new tools. Retain project information on an old project and use this to test the new tool. Always consider possible hidden costs of using something new.

EXAMPLES

Modern Examples

With the wide geographic distribution of manufacturing, the lack of uniformity in methods or tools in projects is no surprise. As long as managers performed and projects were on schedule, no one perceived a need for standardization. The use of the wide area network changed all of that. A major subproject was assigned to evaluate, select, and implement project management methods and tools. Specific tools used included project management software, electronic mail, database management systems, and groupware. To support the standard tools, project templates were created and stored on the network.

Nevertheless, team members continued to show resistance. Compromise carried the team through a transition period. Only new projects were to use the new software.

A lesson learned here is that methods and tools are intimately related to politics, power, and organization. A second lesson learned is that it is prudent to select one site as a pilot for the other remote sites. This approach can be successful if project managers recognize the autonomy and self-interest of each location.

In the case of the construction company, the problem was to extract project information from many incompatible sources to obtain an overall project plan that included all projects and contained resource assignments. Since this had to be carried out on a regular basis each week, the best approach was to write some Visual Basic programs to extract data and standardize it in a spreadsheet. Data was then exported from the spreadsheet to the project management software. This turned out to be easier than retraining all of the project managers at the construction sites.

GUIDELINES

- **Compensate for the inexperience of a project manager.**

  This inexperience tends to be reflected in two major areas. First, the project manager cannot provide perspective on the status and direction of the project, since the project is new to them. Second, the project manager has a difficult time sorting out which issues are important and how issues are related.

Assist a new project manager in developing the plan and setting direction. After that, monitor how the issues in the project are being addressed.

- **Look over all of your tools and methods to see how they fit or conflict.**

Methods and tools conflict in different ways. They can require duplication of effort. They may require you to do manual work to take the results from one tool to another. Different tools may not support the same level of detail.

In such a conflict, often one tool or method will dominate and the other fall into disuse. Which wins? It may be the easiest to use. It may be the one that is politically correct. It is not necessarily the best or most complete. This means that you have to keep a close watch on conflicts as they arise.

- **Base management directives on reality.**

A manager may attend a seminar and become a believer in a certain tool. The manager spreads the word that the tool will be used. But if planning for integration is not performed and follow-up or enforcement is neglected, the tool will not be used. The manager's credibility may be jeopardized.

- **Do not add a new tool or method in the middle of a project; this will generally slow progress.**

If a project extends for more than a few months, it is subject to the "tool of the month" syndrome. This occurs because there is a seller of the new thing. Management may feel that it may make a difference. If someone says to you, "why not try it out?" Beware, there are hidden costs. You first have to divert resources to understand it. Then you have to use it. You have to fit it in with everything else.

- **Test the method first when using a pilot project.**

Pilot projects cannot easily test both methods and tools concurrently. After the method has been tested, then people in the pilot project will have many specific ideas on tool requirements. If a pilot project attempts to test the tool and the method together or the tool alone, then the focus of the pilot is on learning the tool. Guidelines for future use of the tool will be missing, since the method was not validated.

- **Do not ignore gaps in tools.**

A tool gap may occur when you have to manually load data from one tool into another. Another gap occurs when you have to perform manual work because there is no tool. A gap results in more manual effort and is more prone to failure and frustration. This increases the pressure on the team since each team and manager must cope with each and every gap. This can lead to sharp falls in productivity and accuracy of information.

- **Look for interfaces between new tools.**

When a car is modified for a new air conditioner, it is often the case that the team does not think adequately of all of the interfaces with the engine. An example of bad design occurred with a car where the entire air conditioner had to be disassembled to service the engine. How about that—a $500 tuneup? It is the same with tool interfaces. Interfaces include data, procedures, and human interface. Data may not only have to be converted but changed to fit the next tool. People don't want to hear this. They want to test drive the shiny new tool.

- **Assign an expert to every tool.**

Even simple tools require someone who will watch to see that the tool is being used properly. Remember that the job of the expert goes beyond answering questions. It is also to see that the tool is being employed properly and that the expected benefits of the tool are obtained.

- **Allow for the uncertainty inherent in projects rather than adhering to excessive use of formal methods and tools.**

Projects are part science but also an art. Along with the element of uncertainty, projects involve emotions, feelings, and politics. Projects cannot be reduced to a science by quantifying them and then applying formal rules. Quantitative methods have a value, but this value is limited.

- **If someone pitches a new method or tool, only believe that it works when it works in your project.**

  A method or tool depends on the project and its environment for its success. It is working when you see it work, not when people say it is.

- **Gain proficiency in tool or method use through experience rather than through study.**

  Suppose you buy some home accounting software and set it up on your machine. After an hour you get bored. You decide that instead of sitting at the computer, you'll read the manual. That helps, but you will not understand and retain what you are reading unless you go back to the keyboard and try it out. Reading about a method or tool is not as effective as hands-on experience.

- **Look over all of your tools and methods to see how they fit or conflict.**

  Tools and methods can conflict in different ways. They can require duplication of effort; they may require you to do manual work to take the results from one tool to another. The tools may not support the same level of detail.

  When two methods or tools conflict, it is not always true that the best one prevails. The victor may be the easiest to use. It may be the one that is politically correct. Keep an eye on potential areas of conflict and take an active part in resolving the conflict in the way that most benefits the project.

- **When considering a method, think about what type of person can use the method.**

  Any method presumes that the people using the method have certain skills. This applies to basic language skills as well as to complex production systems. If the method requires a star player, and you have few stars, the method is elitist for the small group. Missing skills may mean failed methods.

- **Train close, but not too close, to the time of implementation.**

  If those trained do not use what they have learned, they will gradually forget. Six months later you will have to retrain.

Conversely, if training is done right before a major use, this can cause people to feel pressure.

- **Understand how methods work together.**

  Learning one method is relatively easy, involving only one set of procedures and experts. But when you use several methods, problems occur. You have to rely on documentation and procedures written by different people for different purposes. Be sure to fill in the linkage. Make sure people understand how to move between methods and tools.

  To be effective, use methods consistently and frequently. If you use the method in different ways, you are less likely to develop proficiency and skill, since you are just adapting. House plants are frequent victims of this. People neglect them, then pay too much attention and overwater them.

- **Avoid resource leveling.**

  A method that is supported in many project management systems is resource leveling—don't use it.

  Resource leveling is when the project management software attempts to solve overcommitment of resources by moving tasks around within the boundaries of slack time. This is a good idea and feature in principle but it is not practical. In many systems, you cannot undo the leveling. Watch for this feature in your software and compensate for it. Use manual resource leveling in which you move tasks and changes are reflected in a resource graph.

- **Imposing some tools may mean other subversive tools are used in the organization.**

  Market methods and tools to staff and managers based on appealing to their self-interests. If upper management declares everyone will use a tool, people will pay lip service to the tool, but they may not actually apply it to basic procedures.

- **Avoid overtraining.**

  Overtraining in tools can lead to inertia. You probably have seen this many times in larger organizations. Rather than no training, the organization errs on the side of too much training. People can become so bored and disinterested that any desire that they had to learn the tool is dissipated.

- **Calibrate your tools.**

  Calibration means that the parameters for using the tool and interpreting results are in place. An example of a need for calibration is when you implement a statistical software package that has three levels of difficulty. The proper calibration depends on the level of experience and knowledge available on your team.

- **Focus on results, not process.**

  Do not become so entranced by a method that you forget what its purpose is.

- **Track both effort and elapsed time.**

  This difference is subtle. Elapsed time is typically the duration of the task. Effort is the total number of hours that are being allocated to the task. If no resources are assigned to the task, these are the same in the software. Elapsed time is affected by the calendars and work periods that are employed.

  For example, if three people work 20 hours on a task, the effort is 60 hours. The elapsed time, assuming one shift of work per day, is 2-1/2 days. However, if they work a shift and a half with overtime, then the elapsed time is less than two days. Be sensitive to the effects of changing calendars and schedules when effort and elapsed time are calculated.

- **Use electronic systems in support of project management.**

  If you attempt to use manual methods, the fast pace of the project may outstrip your ability to cope with the project. Also, electronic systems provide visibility of information. Set in your mind what you want the role of the project management software to be. For example, you may want to use it for reporting and for presentation to support analysis.

STATUS CHECK

- Do you have a standard list of methods and tools for project management in your organization?

- How has the list of methods and tools changed over the past few years? Is there a formal method for updating and replacing specific methods and tools?

- Does your organization employ or endorse software tools that are network-based in which sharing of information is supported?

ACTION ITEMS

1. Evaluate your current methods and tools. Begin with the list given earlier in this chapter. In the first column, write each area of project management. Write down the method used in each area in the second column. Write down the tool used in the third column.

   Next, make a list of issues and problems that you have observed with current project management methods and tools. Put these in a row in the first column of a second table. The second column of this table should contain the area of project management from the list. A third column can contain the impact of each problem.

   The first table shows what you have; the second indicates the gaps.

Area of PM	Method	Tool

Issue	Area of PM	Impact

2. Define what tools you would like to have to use with the software your firm already owns. Again, you can use a table. In the first column, list the tool areas. In the second column, give the proposed tool. The third column is for the current tool. The fourth column can contain the benefits of transitioning to the new tools.

3. Estimate the cost and time to implement a new set of tools. Write down answers to the following questions:

   - How many PCs will have to be upgraded for the new tools?

   - Will the network have to be upgraded?

   - How many people will have to be trained in the new methods and tools?

4. In your project defined earlier, identify the methods and tools you would like to use. Compare this to what your organization employs. What advantages are offered by those you would like to use? What are the limitations of the company's methods and tools not selected?

PART II TWO

LIFE SUPPORT FOR YOUR PROJECT— COMMUNICATIONS

CHAPTER 7

PROJECT COMMUNICATIONS

CONTENTS

7

PROJECT COMMUNICATIONS

INTRODUCTION

In a project you communicate throughout the day. You work side by side with team members. You talk with people from other projects. You send and respond to reports, memos, telephone calls, and electronic mail. You can perform good work, direct people well, and show good results, but poor communication within and outside the project team can negate all the things you are doing right.

In this chapter and the next we cover project communications. This chapter centers on types of reports and presentations and provides guidelines for different media forms. The next chapter considers your communications with management on specific issues and activities. This chapter is process-oriented; the next chapter is issues-oriented.

Here is an experience and a lesson learned. A friend was fourth on an agenda for presentation to a board of directors. Each presentation went over its allotted time. Each presentation was slick, featuring fancy graphics slides, some with animation. She sat there with our black-and-white static transparencies and wished she was at home or at the beach—anywhere but in that room.

When her turn came, she had very little time left. Some board members were starting to gather up their belongings so that they could leave. Others were nervous or bored. This was an important presentation that would determine the funding of the project. She decided to cut to the chase. She began with the problem that the project was to face. In this case, it was to create a product that would generate revenue. She outlined the benefits of the proposed product and discussed competition, showed an outline of a project plan, and closed with asking for approval to start. This all took less than ten minutes and we used only four transparencies.

How did it come out? The project was the only one approved. One board member indicated that if the other presenters had spent as much time on the content of the plan as on the fancy graphics, the result might have been different for them.

The lesson learned was to focus on content more than on format.

PURPOSE AND SCOPE

Many people take communications and presentations to extremes. They spend too much time worrying and preparing. Others just walk into a room and ad lib. The purpose of this chapter is to help you take a balanced approach to communications. Take each contact, presentation, or telephone call seriously. Be ready to address almost any aspect of the project at a moment's notice. On the other hand, have some enjoyment with the project experiences and lessons learned.

Conveying the message is very important in project communications. This goes beyond getting a single message across one time. Project communications involve building and maintaining communications paths with many different people, since projects involve team activities and decisions. Communications are also useful in gathering information and resolving problems more effectively.

To reach the goal of a balanced approach to communications, you will define each step in the communications process, considering almost all forms of communications, formal and informal.

The scope of this chapter includes communications in various types of reports and presentations.

APPROACH

Always be ready to communicate effectively. Beyond the ability to address issues and crises, this is one of the most important attributes of a good project leader. If your demeanor is sour or down, then the audience may interpret this as indicating a problem in the project. It is all right to show anger, concern, or worry if that helps in getting an issue resolved or in advancing the project.

ALTERNATIVE COMMUNICATIONS MEDIA

Communications have become more complex because you communicate with more people than people did in the past. Also, you have more choices of the medium of communications. Here is a short list of the possibilities, with specific guidelines on using each one:

- **In-person informal communications**

    This is the best for discussing something or gathering information. You can get a better overall impression of what is going on if you go to a person's office rather than running into someone in the hallway. Many different topics can be explored in this casual atmosphere. You get more of the person's attention without interruptions. When you stop by, indicate why you are there, what information you need, and how much of the person's time you need.

    If someone contacts you, be ready with a smile and look the person in the eye. Always be ready for any informal visit. You might even enter your own office as if you are a visitor and see what impression you are conveying.

    Whether you are the sender or the receiver, check facial expressions. Does the person you are talking with appear closed up by having legs or arms crossed? Do either of you appear nervous? Scan what is on a person's desk. It will tell you what the person's priorities are.

- **Formal meetings**

    This is a chance to cover topics with a larger audience in a structured format. More ground can be covered. There is an opportunity to gain consensus. Formal meetings require more preparation and thought.

- **Telephone**

    This is a good medium for following up on specific nonpolitical points if the person is remote from you.

    If you are on the receiving end of a call, answer with a greeting and your name. Let the caller do the talking. Just listen. Try to detect from the caller's tone whether he is nervous, upset, or angry.

    Maintain a telephone log of all calls. This will help jog your memory later. If a caller makes a specific request that impacts the project, ask for the request in writing so that it can become part of the project file, since it could affect the work or the schedule.

- **Telephone contact with an intermediary**

  Careful here—any message you convey is subject to misinterpretation. Stick to a straightforward message that is clear and unambiguous. Organize your thoughts before you call.

  After identifying yourself on the phone, establish rapport with the person answering. Then move to the message. Imagine that you are writing down the message along with the intermediary and speak at a pace that allows this.

  If a secretary or assistant calls you on behalf of someone else, get the details, then repeat the message back to ensure that you have received it correctly. If you are asked to come to a meeting, do some research. How big is the room? Which department controls the scheduling of the room? The answers to these questions can reveal who the other audience members are and tip you off as to what may be covered.

- **Voice mail**

  Keep the message brief. Make notes before you call, writing down the subject and key points. If the call is for information only and needs no response, indicate this. For your own voice mail line, keep the greeting short and avoid being cute, which may distract the caller.

- **Pagers**

  Adopt a code system for each person you will be paging frequently. For example, if the need for contact is urgent, enter 911. If the need is informational, enter 411. Returning pages promptly indicates to the sender that you treat the business seriously.

- **Facsimile**

  Try to get as much on the cover sheet as you can. Try to use electronic mail instead of faxes since it is more private. If the person being faxed is not located near the machine, call and alert the person that you are sending a fax.

Fax the message yourself to ensure that it was sent. Someone else who doesn't really care might insert it into the machine and walk off. The fax may fail. Someone else walks up and removes your pages and leaves. Finding the pages neatly stacked, you mistakenly assume that the fax went through successfully. Also, watch the time of day. If you fax at lunchtime, it's possible no one will be there to pick it up. If you fax at busy times and are able to get through, then your fax may end up on the back of someone else's. If you are going to send copies to several people, address each person on a separate sheet. Always hand-sign the cover sheet.

- **Electronic mail**

  If you are going to send a lengthy e-mail message, write it first in a word processor, since this is a much better editor than the e-mail text editor. Also, many e-mail systems do not have a spell-checker.

  Make sure the subject line is short and clear. Include any issue or topic that is the focus of the message.

  Establish group mailing lists in the e-mail system for the inner project team and outer project team. This will save you from having to type in all of the e-mail addresses each time.

- **The Web and Internet**

  The intranets, the Internet, and the World Wide Web have so much useful information for projects that being selective is the best skill to acquire. Don't flood the receivers with too much information.

  If you are going to post documents or information on web pages, then concentrate on content. Don't spend a lot of time on making it cute by customizing HTML code. Cute stuff can backfire since the audience may assume that you have nothing better to do. If you are going to establish a web site, maintain the site with up-to-date information or go back to e-mail. Remember that a key duty will be maintaining any and all Web information you establish.

- **Groupware**

  Some people employ groupware like electronic mail. However, groupware has the advantage of allowing several people to comment or to build documents together. In using groupware, stick to the basic features. If you use exotic features, other people may not be able to participate fully.

- **Videoconferencing**

  This is one of the more abused media forms, even while in limited use. To cope with limited communications capacity, only a limited number of frames are transmitted per second. This means that the image is jagged. Use the videoconference to show charts, tables, photos, or drawings. You can then communicate by voice or text in discussing the figures.

  This is a communications form that has a great deal of potential for some applications. For example, in engineering and manufacturing, the technician on the floor can ask questions of a design engineer regarding a drawing. For construction companies, videoconferencing is one of the best electronic tools available, since it puts multiple construction sites in touch with each other and with headquarters.

- **Memoranda and letters**

  Keep written memoranda and letters short and to the point. Adopt a simple writing style. Avoid long sentences and complex terms. Make sure the subject is short and to the point. Think about what follow-up measures you will take to ensure that the message got through.

- **Reports**

  When writing the report, keep in mind who the audience is and keep the focus narrowed to this. Print on one side of a piece of paper. Although this consumes more paper, the document will be easier to follow. Number and date each page with the title so that it can be reassembled if separated. Always assume that any written communication will be copied and distributed.

MESSAGE FAILURE AND SUCCESS

What constitutes success in project communications? The message not only got through to the receiver, but also you received action or a response.

What constitutes failure in project communications? The message was misinterpreted or ignored. The failure is always the fault of the sender, who is responsible for all steps in preparing, sending, and following up on the communication.

SIX STEPS IN COMMUNICATIONS

The following six steps were standard practice on the railroads. With most communications being telegraphed, senders had to reduce their communications to shorthand and code. Follow these six steps in communications to ensure success:

Step 1: Define the Purpose and Audience

Whom do you wish to reach? Think about the person or people you are trying to reach. What do the people think of you and the project? What is their attitude? What is their knowledge of the project? What can you assume they know so that the communication can be shorter? What are they likely to do after they receive the message? To whom will they pass the communication? Answering these questions will help you to determine the degree of detail and background required, set the tone of the communication, and determine what medium is most appropriate.

Why do you wish to communicate? What would happen if you did not communicate now? With the overload of information in many organizations, treat any of your communications as if it costs great money and effort. If you are in doubt and the reasons for the communication are unclear, then wait. Is the reason for communicating complex? That is, do you need to address a number of issues? Consider breaking down the communication into addressing one person at a time or addressing smaller groups.

Step 2: Form the Message

The message is not the communication. The message is contained in the communication. Based on your skill, the receiver may or may not be able to decipher the message from the communication. To avoid confusion, first construct an outline of what you want to say. In general, your outline should include the following:

1. An introduction that identifies the problem or situation
2. The detailed steps of the message (who, what, when, where, how)
3. The desired action of the recipient
4. Expected feedback from the message

Example: Manufacturing

The project manager at the manufacturing firm had to request a detailed blueprint of the offices in Singapore in order to lay out the network and estimate costs. However, a blueprint can mean different things to different people. What if some vague diagram were provided? It would be worthless and the information request would have to be repeated. The communications outline first listed the request and reasons behind it. This was followed by the date required and what was to be done with the information. The last item was a sample blueprint of the headquarters building as an example. This was appropriate and provided the necessary example along with the information on the request.

Step 3: Determine the Medium and Timing

Confine all discussion of issues to in-person contact or direct telephone contact. You want to have direct interaction with the person so that any questions can be answered. Use electronic mail or groupware for routine messages. Avoid facsimile because of the many possibilities for missed communications. Keep communications informal. Use formal meetings and settings as a backup for escalation and for general impressions.

After you select the medium for the message, think about how the receiver will respond and what media will be used. Assume that the messages will fly back and forth.

Timing is important if you want to get someone's attention. When should you send the message? The answer, as you saw in the discussion of various media forms, depends on the medium, the objective, the audience, and the message.

Step 4: Formulate the Communication

In this step you now package the message inside the communication. Whether you are dealing with verbal or written media, if it is important, expand your outline and build the communications. If the communication is verbal, create a series of bullet items. If the communication is written, prepare the document. Write with words of ten letters or less whenever possible. Write in simple sentences, usually no longer than 10 words, and form short, succinct paragraphs. Avoid jargon—especially project management jargon, such as *critical path*, *PERT*, *GANTT*, and *critical resources*.

Step 5: Deliver the Communication

If the delivery of your communication brings many questions from the recipient, suggest that you go over the message with the person in a face-to-face situation. If complex questions arise during direct contact, set another meeting to resolve all the issues raised.

Step 6: Follow Up on the Communication

If you send messages and you fail to follow up, people may think that the message was not important.

First, make a note in a log as to what you sent and when. Track whether you receive any feedback or response. Plan ahead for follow-up on whether the message was received and what actions are flowing from it. In most cases involving politically sensitive topics, the only evident response may be acknowledgment of receipt of the message, since they will need time to think about a response.

Figure 7.1: Sample Form for Communications

No.: _____ Date: _____

Title: _____

Audience: _____

Purpose: _____

Message:_____

_____

_____

_____

_____

_____

Media Selected: _____

Expected Action: _____

Expected Time of Action: _____

Actual Time of Action: _____

Real Action that Occurred: _____

Lessons Learned: _____

_____

Notes:

Assign a number to each form so that you can track it later.

File these by subject or by date.

Obviously, to do this for many messages is absurd and impossible. This is another reason for not sending out many messages. If you have to send more messages in total than you can pursue, identify the critical messages and follow up on them. Place messages on issues, budget, schedule, and important resource topics in this category.

Figure 7.1 gives elements to use to track and improve communications skills.

EFFECTIVE REPORTS AND PRESENTATIONS

When you think about project presentations and reports, your mind will often turn to technical details and in-depth discussions of issues. Most people want to plunge in and make a list of details and then formulate a report or presentation. This is unwise. Here are six topics which must be considered in creating an effective report or presentation.

Six Composition Decisions

1. **Medium or format**   What is the medium of the message?

2. **Length**   How long will the communication be?

3. **Organization**   Where will you start your presentation? Where will it end? How will you get from the start to the end (order of presentation)?

4. **Method of argument**   Will you use project data, your authority or experience, or project history? What will be the basis for your support?

5. **Attitude toward audience**   What is the attitude you wish to convey toward your audience (friendly, hostile, polite, informal, etc.)?

6. **Impression**   After your message is heard or read, what do you want the audience to think of you?

With this as your strategy, you can begin to assemble information, tables, graphs, plans, etc. This is the evidence for your presentation.

Presentation Style

Next, choose one of three presentation styles:

1. **Descriptive Report and Presentation**   An issue, situation, or opportunity can be described to an audience. You begin with an introduction to the subject. Next, establish your credibility. Why are you qualified to talk about this? Here you might cite project experience. Third, give an overview of the topic. This is followed by the details on the subject. Now bring the audience back to the initial topic so that they can relate the overview and detail to the topic.

   This style works for a travel show or for some academic presentations but has limited effectiveness in project management. In projects you are often trying to gain support or approval. Description often is too general, leaving too many loose ends.

2. **Analytical Report and Presentation**   Here you might be reporting on an issue in a project. You start with identifying what you are analyzing. First provide your credentials. The question addressed next is "What is the current state of affairs with the issue?" You can move from general to specific and back to general, as in the descriptive presentation. Now identify the methods and tools employed in the analysis. You end the presentation with conclusions and recommendations.

   This style is good for milestone assessment and other evaluation type work, but is not well suited to the major presentations of the project.

3. **Persuasive Report and Presentation**   Concentrate your effort and practice in this presentation style. As with the other two styles, you begin with an introduction and qualifications. Next, what is the need? What are you trying to address? Avoid the solution or benefits. After answering these questions, answer the question, "What will happen if the need is not addressed?" Point out grim and unwelcome consequences. Then change the tone to one of optimism and talk about what will happen if the need is addressed. What benefits will accrue? You have warmed up your audience and prepared them well. Now move to the solution.

MEETINGS

Some general suggestions regarding meetings are as follows:

- Do a great deal of preparation for meetings. Collect agenda items for meetings in advance.
- Actively run the meeting. Keep meetings focused on an agenda.
- Minimize meetings due to the effort and the impact of lost time on the project.
- Meet to discuss lessons learned rather than to discuss status.

Here are some specific comments on two types of project meetings:

- **Project Kickoff Meeting** You are starting the project. The meeting will introduce members of the project team to each other and set the stage for the project. Have people introduce themselves and explain what experience and expertise they bring to the table. This is especially important since you are establishing the tone for the project in this first meeting.

   Prepare for the meeting by developing the project plan from the template, but leave out the lowest level of detail. Delegate that task to the team members. Set down the ground rules for communications, reporting, work, and issues. Make this as structured as possible. If you start out vague, then you have damaged the project at the onset. To build the group into a team, develop the initial list of issues and detailed tasks as a team.

- **Milestone Meeting** This is a meeting where a milestone or end product is presented and reviewed. As the meeting begins, provide the audience with checklists of questions and guidelines for evaluation. Let the people know what you expect to get out of the meeting. What actions are possible? As you go through the review of the end product, make notes on an easel or board as to what issues and questions are raised. At the end of the review, start going through the list on the board and either get closure or assign topics out for analysis.

- **Issues Meeting** Identify the two or three issues to be addressed and who should attend for each issue. Set strict time parameters for each issue and invite only those needed to each issue discussion. Identify action items for each issue and decide how to follow up on these.

- **Lessons Learned Meeting**   These meetings can be based on achieving a specific milestone. You would ask what people learned in doing the tasks that led to the milestone. Another option for this kind of meeting is to take a time period and ask for lessons learned for that period. In either case, start with a general discussion. Then identify the following elements and record them so that others can benefit:

  —How the lesson learned can be generalized

  —How someone would use the lesson learned in practice

  —The benefits of the lesson learned

  —Who to contact if a person has questions about the lesson learned

  —How to add further detail later to the lesson learned

  —How to measure the results of the lesson learned

EXAMPLES

The Railroad Example

Communications on the railroads were divided into several areas. Direct contact took place at the work sites. Results from these communications would progress into the form of telegraph messages and paper to headquarters. Since many people could not write, a designated person took down what was said and converted it into a shorthand code. At headquarters, railroad construction was so complex that the modern organization emerged in dividing logistics, accounting and finance, and other areas. Meetings and conference rooms were common. Agents and managers were assigned to go out in the field and make site inspections. They would relay their findings in code to ensure privacy.

Modern Examples

Communications on the project at the manufacturing firm started with electronic mail and faxes. Problems arose because the faxes would get out of sync. While preparing a response to the first fax, a second fax would come in. These faxes were mixed in with standard daily work. Faxes were lost. Electronic mail growth was slow and uneven.

The project manager stepped in to help by supporting electronic mail use. Once e-mail was established, usage soared.

The lesson learned here was that these events could have been foreseen and planned for in advance. The project manager had to react to each situation. Some of the problems could have been prevented with better planning and organization.

For the engineering and construction firm, the problem was that people were used to a variety of different communications media. No standardized pattern existed. This created communications problems because people employed their own style between written, voice, and electronic media.

The project manager was able to standardize the distribution of project plans and issue management. Extending standardization any further would have consumed too much effort and engendered too much hostility.

GUIDELINES

- **Think of the self-interest of the audience when you supply project information.**

  Remember that the importance of a project lies in the eyes of the beholder.

- **Avoid secondhand communications.**

  Secondhand communications means going through intermediaries. You know from the elementary school game of "telephone" how messages get changed as they pass from person to person. Also, it is highly unlikely that the intermediary will convey the passion, interest, or other emotion that accompanied the message. This lowers the likelihood of the message leading to success.

- **Democracy in a project has its place, but so does autocracy.**

  In a collaborative environment for a project you seek to encourage greater communications about issues, lessons learned, and project status. However, also maintain order in the project and stomp out any rumor mills. This is a challenge between democracy and authority for the project leader.

- **Detect indirect resistance by observation.**

  When people communicate with or respond to you, they convey their feelings and attitudes. You should be able to detect resistance on specific topics. Maintain frequent one-on-one contact to detect changes or nuances in demeanor.

- **Manage small leaks of information while they are small.**

  Most project leaders can cope with leaks of information about the project to people outside of the project. The basic problem is that what might be an annoyance in the project becomes a major crisis through the retelling of the information again and again by different people. Coping with substantial disinformation is time-consuming. If possible, determine the source of the leaks and discuss the issue with that person.

- **Be consistent in what you relate to each of the members of the project team.**

  Assume that everything you tell someone on the team is known to the entire team in less than one hour.

- **Operate on the assumption that you might create an enemy in each meeting.**

  Enemies (or friends) are made during the process of communication. Once you have made an enemy, it will be difficult, if not impossible, to undo. Be considerate and avoid unnecessary criticism of others throughout the communication process.

- **Keep the volume of written memos among project members as low as possible.**

  The volume of memos is usually inversely related to progress. Experience indicates that memoranda volume increases when the project is under stress. People naturally want to cover themselves and appear busy. If people are working productively on the tasks, they have little time for memos.

• **Confine your meeting time to issues.**

Project issues are often more interesting than project status. People tend to want to get involved and put their fingerprint on issues. People are bored by status.

• **Use simple language and avoid arcane jargon.**

Using unfamiliar or arcane words was encouraged at one time as a demonstration of intelligence. Now it is viewed negatively because people miss the meaning of what you are saying. The communications path to a person consists of first understanding the language, then understanding the words, and then deciphering the meaning of what is said. If people can't understand you, your communications are not going to be effective.

• **Vary the project report formats.**

If you continually use the same format and structure for reports, you will not have an impact on your audience. Your goal is to obtain a decision and an understanding, but the message may be blocked by the format of the project reports. Design a series of formats—one for use with routine reporting, another series for analysis, and a third series for decision-making.

• **Share project information.**

Sharing project information builds trust. People often hold project information close to the vest. They think that if it gets disclosed, their position will weaken. This occurs more often when professional schedulers are tracking progress for management. It is always better to share information on the project. You don't have to share the political perspective or details about issues in progress, but be open about status and activities.

• **In order to raise morale, look at the worst that can happen.**

In communications, when you encounter a difficult issue or crisis, take the time to examine the worse case scenario. Often, this is not as bad as people may have thought.

• **Have people leave their present agendas and schemes outside before a meeting.**

As a project leader, you will have to address these agendas directly at the beginning of the meeting. It is better to tell people to leave their "guns" outside and see you privately at a later time.

- **Try to sit in on meetings for different projects and teams.**

Learn something about style and what works and does not work, no matter what the content might be. Focus on the flow of the meeting, the interaction of the people, and the structure of the meeting.

- **Keep project meetings to no more than an hour.**

Meetings that last more than an hour tend to generate more heat than light. Extended project meetings tend to disrupt other work. People may get too worked up about an issue during the meeting and productivity will then drop for some time after the meeting.

- **Provide a forum to encourage the sharing of lessons learned.**

If you do not, you will suffer the penalty of repeating the lessons. Having the team members share their experiences provides reinforcement and support. Otherwise, a team member will have to go through the same processes as other team members without the benefit of the experiences and lessons learned by the others.

- **Vary project meeting dates and times to increase the level of awareness.**

Periodic meetings are often preferred by people who like routine. But holding periodic project meetings will lull a project team into complacency. Also, issues do not conveniently mature and become ready for resolution on the same schedule. Drop the periodic meetings. Make the next meeting "to be announced" and notify people several days in advance.

- **Stick to simple visual aids and have handouts in case these fail.**

The more you depend on extensive audio-visual aids, the more likely they will break. Exotic visual aids include electronic CRT screen projectors, nonstandard slide projectors, and overhead projectors without substantial fans. Do not trade content for slickness. Do not rely on equipment that has no spares or on-site support.

- **Evaluate yourself after a presentation.**

    After a presentation there is a tendency to want to forget everything and go on to something you like to do. But first, sit down and be your own worst critic. Did you achieve the results you were after? What did the audience do in the meeting? How did they react to you? How did they react to other speakers?

- **Consider chart appearance.**

    Charts can be confusing, even if the information is valuable and correct. The choice of colors, shadings, format, lettering, fonts for letters, and wording are all important. Create your presentation and set it aside for some time. Then go back and shuffle the presentation order. Pick up a chart at random and see if you can understand it. In a project presentation, discuss the impact of the chart—not the detailed meaning of the chart.

- **Hand out all materials at the start of the presentation to make life easier for the audience.**

    The old argument was that people would read ahead of your presentation and would lose interest. However, people get more from the information if they have the materials in front of them. Also, handing out materials at the beginning of the presentation minimizes surprises.

- **Hold informal meetings more often than formal.**

    The more formal meetings you have, the more people will become involved in the issue. With more heat and attention, many people are likely to defer action. It is best if possible to resolve issues and get decisions informally and with a low profile. The decision can later be announced formally. Also, informal meetings often convey more information than formal meetings. Formal meetings tend to have a rigid agenda with less time for questions. The presentation tends to be more rigid in terms of overheads, slides, and handouts. In informal meetings you can get questions and issues out more easily since there is less structure to the meeting. One good strategy is to have an informal pre-meeting to solicit issues and questions to be covered later at a formal meeting.

STATUS CHECK

- What is the level of your awareness of being able to differentiate between the message and the overall communication?
- Think through the communication process you use today. How often do you have to clarify messages that were not properly received?
- How much project time is spent on determining status? How much is spent on issues?
- Do you have any standard guidelines for presentations to management? Do these fit all of the types of projects in your organization?

ACTION ITEMS

1. Go into your project file and grab several memos or copies of electronic mail that you generated. Review these by asking the following questions:

    Does the communication fit the audience?

    Can you discern the message through the communication?

    What was the result of the communication? When did it happen?

2. Start keeping a log of your communications. Note the date and time of contact, the person contacted, the nature of contact, the response, the date and time of response, and the action that resulted.

3. Develop an approach for modifying your method of communicating. Start with one media form at a time. Work on electronic mail and faxes, since these are relatively short and focused.

CD-ROM ITEMS

Use these company example files to help in communications.

07-01   Rylande Corporation

07-02   Rylande Project Template for Process Improvement

07-03   Rylande Project Plan

CHAPTER 8

GETTING AND KEEPING
MANAGEMENT SUPPORT

CONTENTS

8

GETTING AND KEEPING
MANAGEMENT SUPPORT

INTRODUCTION

While you can recover multiple times from problems and failure within the project team, you get very few chances with management. If you are successful in your first major presentation, the favorable impression you create will last a long time. However, the margin for error is small. If you don't pay attention to management communications and their nuances, you could undo all of your other good work.

Example

The project leader of the consumer products team almost destroyed the project during his first presentation. He indicated to management that the project would replace all of the schedulers and scheduling. Several managers winced and became confused. They were not aware of what schedulers did in the first place and the plan sounded drastic. The meeting ended without approval. A misimpression was created that the project was too revolutionary. It took two months of behind-the-scenes marketing to correct this before the project idea could be presented again.

PURPOSE AND SCOPE

This chapter focuses on key events in your management contacts during the life cycle of a typical project. The previous chapter gave general guidelines on communications; in this chapter more specific, pointed suggestions are given. The purpose is to help you win support. This chapter provides practical advice as well as tips on how to avoid failure.

The scope of this chapter begins with getting the project idea approved and moves through the completion or termination of the project. Both formal and informal communications with management are included.

APPROACH

The traditional method in project management is that as you approach a major event or milestone in a project, you put together a formal presentation to management. The presentation then may or may not occur. If it does, some follow-up may be required, but the attention returns to the work. The contact with management is a temporary event, perhaps viewed as an interruption in the process.

This concept of communications with management is fundamentally wrong. Instead, consider management communications as an integral part of overall communications. Informing and working with management on a continuous basis are major roles for the project manager which are just as important as obtaining status and addressing issues.

The discussion of this better approach to management communications is divided into informal and formal communications. The most effective strategy is to use informal methods as a basis for communications and smoothly and continuously build up to formal presentations. The formal presentations will then be followed up on by information communications.

About the Audience

The old school of thought was to keep your manager informed and that would be sufficient. However, managers come and go. Also, the elapsed time of the project may be such that the manager may move to another position. A suggested alternative is to identify and keep informed a set of three to four managers in different parts of the organization. These people will be your direct audience for both informal and formal communications.

With which managers should you develop a rapport? Choose a combination of general, high level managers and line managers who are interested in the outcome of the project, or who are supplying people to the project. Make a list of several managers of each type to call on if you need backup.

This method has several benefits. First, you have a plan in place for backup if one manager leaves. You also have continuity, since the existing managers can assist in updating the new manager as well. A third benefit is that you have a wider audience to give you feedback before formal presentations.

INFORMAL COMMUNICATIONS

Benefits of Informal Communications

Here are some benefits of informal communications:

- The managers can give you their reactions to a presentation informally prior to formal presentation. People tend to be more open in a one-on-one situation.

- You provide others with information on issues. They can then take the information and work the issue for you behind the scenes. It is often best to solve politically sensitive issues "off-line."

- Status information can arm others to answer any questions or concerns about the project. This prevents both defensiveness when someone questions the project and the need to call you to clarify a point.

What do these benefits add up to? You become more proactive. You are getting information out to people. When formal presentations occur, they are almost anticlimactic, since several members of the audience already know what you are going to say. This also means less chance of a surprise.

How to Make Contact Informally

Here are some pointers on informal contacts with management.

- **How to contact managers** Plan on casually running into several managers each week. Planning and "casual" contact appear contradictory, but they are not. Plan how you can informally contact managers in the hallway, copier room, or their offices. Study their work patterns. Usually, the best time to run into people or stop by their office is early in the morning.

- **Extent of contact**   Plan on no more than five minutes of total contact, unless the manager indicates that he wants to spend more time with you.

- **What to cover in the contact**   Always start with status. Let the manager know some good news first. This is a positive way to start the day. If you want to discuss or present an issue, gradually lead into it. Starting with the issue is too negative.

  If you desire feedback on the major points of a formal presentation, reveal some of the key parts of the presentation for reaction. You get not only the manager's reaction but also the manager's understanding. This typically means that the manager can provide assistance during the presentation.

- **A manager's concerns and comments**   The previous point stressed the transfer of information from you to the manager. It works both ways. The manager may hear something that impacts your project. The informal contact allows the manager a chance to inform you without a record in writing.

- **A manager's ideas**   Incorporate the ideas suggested by the managers and then provide them with feedback by showing how the presentation or report changed after their input.

With this amount of contact, you and the project are quite visible to certain managers. However, to others you are unknown until there is a formal presentation. This is desirable. If you receive a great deal of management attention openly, people will become jealous and may take shots at the project. To keep a low profile, favor a structure with extensive informal contacts and little obvious visibility.

FORMAL PRESENTATIONS

Before the Presentation

Whether it be a report or an oral presentation, try these guidelines:

- Keep all materials in a draft form. Label them as a draft.

  This will give people the impression that they can have input prior to the final form, as well as providing you with the ability to improve and make changes. The more people are involved, the more feedback you will get and the more buy-in you will receive.

- For verbal presentations, ask line managers of the project team members to be present along with team members. Plan to spread the credit around.

- Use successive dry runs to keep improving the material.

During the Presentation

Here are some tips to consider in making the presentation:

- Minimize the number of charts. Too many can be confusing.
- Hand out all charts at the beginning. This allows the audience to see the entire presentation and surprises are avoided.
- All charts should express complete thoughts and sentences. If you use lists without additional information, people can misunderstand what is going on.
- Encourage feedback and questions. Follow up on each item either in the meeting or shortly afterward.
- Make sure you have dangling items so that you have the opportunity to follow up afterwards for more marketing.
- Walk around as you give the presentation and be animated.

After the Presentation

Follow up after the presentation with managers who were in the audience and further explain any points that were unclear. Asking what they thought is too direct and shows a lack of confidence. If a manager has an opinion, he will express it.

Immediately after the presentation, grade yourself by using the following checklist:

Presentation Evaluation

- **Material**

  How well was it organized?

  Was the material relevant to the theme of the presentation?

  Were people able to understand it easily? Did they ask what terms meant?

  Was there too much material?

  Did you find that you lacked material to respond to specific issues?

- **Presentation style**

  Were you too formal?

  Did you receive many questions or comments?

  How did you respond to comments?

  Did you read from the charts?

- **Audience**

  Who attended the presentation?

  Were the key players there?

  Was the audience attentive? Were there any interruptions?

After a presentation, take the time to write down the answers to the above questions and file this evaluation for reference. If needed, refine the presentation to the way it should have been.

MARKETING

Critical Marketing Milestones

Almost all projects, even the smallest, have key milestones and end products. Here are the ones to be considered in detail.

Marketing Milestones

- Marketing the project idea
- Marketing the project plan

- Communicating project status to management

- Presenting and addressing an issue or opportunity

- Obtaining management decisions

- Coping with poor management decisions

- Taking action after a decision

- Changing a project

- Terminating a project

For each of these let's examine the background, purpose, approach, tactical suggestions and hints, and how to assess yourself after the presentation.

Marketing the Project Idea

- **Background**

  Project ideas don't just surface on their own. Someone has the idea and either that person or somebody else follows up by becoming the champion for the plan concept. The fate of the idea is closely linked with the person who is pushing for approval of the idea.

- **Purpose**

  The aim is to gain approval from a manager to develop a project plan and determine the initial feasibility. This is positive, since it gives management an inexpensive way to assess the will of the person who is pushing for the project.

- **Approach**

  The favored approach is to build support for the project gradually. Sell individual managers one-on-one. Show how the project concept fits in with their own interests. Avoid a pitch based on vague terms and concepts.

  To generate enthusiasm as well as gain approval, show the benefits of your plan. For example, a project leader in one of the examples did this for her network by pointing to how issues could be resolved through a network. This was much more effective than trying to sell dry network concepts.

- **Suggestions and hints**

  Try to have managers adopt the project idea as their own and then act as apostles in marketing it to other managers you cannot reach yourself. Your plan is to build a set of cadres to support and sponsor the idea. Keep the idea verbal so that it is flexible. Once you put it in writing, you will tend to become locked in.

- **Measuring your performance**

  The obvious measure is the answer to the question, "Was the concept approved for planning?" But this is not enough. Also assess the degree of enthusiasm and excitement for the project.

  Another test is to ask managers what their impression is of the project.

Marketing the Project Plan

- **Background**

  It is not enough to gain approval of the project plan; you must obtain resources as well, garnering political support to get the necessary funding.

- **Purpose**

  You should obtain the following after initial approval of the project plan: support for providing resources to the project for the first stage of work, interest in the project for continued contact, and management input on the plan.

- **Approach**

  One technique is to reveal the overall, high-level plan to managers. Show the major task areas, general dates, and dependencies. Then refine the plan and come back with more detail, as well as resources needed. It is too much to expect people to grasp all of the detail the first time. Presell the project top down and bottom up at the same time. Focus on the people who will benefit from the successful completion of the project.

- **Suggestions and hints**

Incremental marketing is the key. This gives you a further opportunity for management contact and contact with the future beneficiaries of the project. Make it appear in form if not in reality that the development of the plan is a team effort. Make sure that you give people credit for their input.

- **Measuring your performance**

Did you obtain approval?

Do people have a clear idea of the project and are they supportive?

Did you obtain resources to get started?

Communicating Project Status to Management

- **Background**

Conveying status information is not just walking up and saying the project is going along well or writing a memo to that effect. Strive to continue to build rapport and strengthen support. Treat the supplying of status information as a continuous process rather than a periodic activity you do once a week or once a month.

- **Purpose**

The basic purpose is to enlist and build support for the project. A more immediate purpose is to convey an understanding of the project status. Another purpose is to pave the way for the resolution of issues.

- **Approach**

Provide status in informal one-on-one meetings. Start with the overall state of the project and then zoom in on a detailed issue or specific milestone. Then relate an interesting war story or experience. Alert managers to a looming issue so that they can prepare for it.

- **Suggestions and hints**

  Develop a version of the status of the project and what you want to say each morning. Rehearse this informally with a member of the project team. Then set out on your mission to relate the status to one or two managers. Do this several times a week. Managers you contact will become involved and more interested. They will look forward to your contacts.

- **Measuring your performance**

  How many people have you contacted this week?

  How has your relationship grown with them since you started?

  Are they more interested in the project now?

Presenting an Issue

- **Background**

  Present the issue or opportunity along with alternative actions, and get a decision. This way, you are presenting solutions as well as bringing up problems.

- **Purpose**

  The purpose of the presentation of an issue is for management to understand the impact of the issue, why action is necessary now, and the suggested decision, actions, and anticipated results. Avoid getting bogged down in the details of the issue.

- **Approach**

  The first stage is to alert managers through informal communications that an issue is coming. This will mitigate any feelings of surprise. Next, follow the steps for issues suggested later in the book in Chapter 18. When you have the materials ready, present a complete picture.

- **Suggestions and hints**

  Don't cry wolf over an issue. You can alert management and indicate that you are tracking it. Let it mature. Carefully plan the sequence of issues and opportunities that will be presented to management. Insert positive opportunities in between issues to avoid leaving a lingering negative impression.

- **Measuring your performance**

The bottom line is whether you receive approval for the decision and actions you proposed.

More than that, however, did you establish a positive pattern of managing issues?

Obtaining Management Decisions

- **Background**

Many immediate management decisions after a presentation will be negative. To deal with this, during the presentation indicate what you will do while the managers ponder the presentation. In that way, you can continue the project. The managers will know that work is still going on and will feel less pressure to make a decision.

- **Purpose**

The purpose is to obtain a decision, but as important as a decision is, your ultimate goal is support.

- **Approach**

Presell the decision through the informal contacts. Informal approval is easier to obtain than formal. Avoid a formal memo of the decision if you can. The formal presentation can stress the actions that flow from the decision, rather than the decision itself.

- **Suggestions and hints**

A basic suggestion is to indicate that a decision will be needed way in advance. Then show that this decision does not bear any significant risk for management. Instead, focus on the benefits that will flow from the actions. This turns the spotlight away from the decision. As you approach the decision, move the attention to actions—again moving the focus from the decision. This will make the decision more natural.

- **Measuring your performance**

  Did you get the decision you wanted when you required it? How much good will did the decision cost? How hard did you have to sell the decision?

Coping with Poor Management Decisions

- **Background**

  You wanted a management decision and you got the wrong one. They did not approve the requested resources or the budget. What do you do?

- **Purpose**

  The purpose is not to go back in and reverse the decision. What you seek to do, rather, is to mitigate the effects of the decision on the project. It is most important to keep up progress and momentum in the project. Do not let a cloud of doom hang over the project.

- **Approach**

  First, analyze why the undesirable decision was made. What is the difference between your perception of the situation and management's? Next, look at the impact of the management decision—immediate and long-term. How can you counter the effects of the decision to protect the project and keep up the work?

- **Suggestions and hints**

  Expect that poor decisions will be made. This often happens through a misunderstanding. Once key managers have taken public stances that cannot be reversed, figure out how to work informally behind the scenes to counteract the negative effects of the decision.

- **Measuring your performance**

  Were you able to control the damage from the decision?

Have you determined how to go back to management with pieces of information to get some change in the decision?

Taking Action After the Decision

- **Background**

  Actions flow from a decision. If no actions follow a decision, the decision is likely to have little meaning. If the gap between the decision and the subsequent actions is too big, people will not be able to understand and relate to the actions.

- **Purpose**

  The goal is to implement actions immediately after decisions are made.

- **Approach**

  Include the actions in your presentation. Link the actions to the situation by explaining what benefits will likely follow. Then you can back into the decision.

- **Suggestions and hints**

  Make sure that you have a complete list of actions ready to go. Actions can be policies and procedures, as well as resource actions.

- **Measuring your performance:**

  Were the actions implemented?

  Do people clearly see the connection between the decisions and the actions?

Changing a Project

- **Background**

  On many projects lasting six months or more, you will be faced with selling project change to management. This occurs naturally for several reasons. For example, change external to the project may affect the project. Also, knowledge gained from the project team can be employed to change the project.

- **Purpose**

  The goal is to accomplish the change in the project while maintaining the confidence of management in the project.

- **Approach**

  Bundle all possible changes into the marketing of the change. If you change the project many times, management will lose confidence in you and the project. Managers might think that things are out of control.

  Indicate why the change is needed by explaining what will happen if things continue as they are. Then move to the benefits of the change.

- **Suggestions and hints**

  Point out at the start of the project that the project could change due to events. Keep alerting people to this possibility. Then, as you approach the change, give attention to some of the different problems that could be solved by change. Indicate that you are packaging the solutions into a major change and that after this the project will move into a period of stability.

- **Measuring your performance**

  More important than getting the change approved, do you still have management's support and confidence?

Terminating a Project

- **Background**

  This is unpleasant but necessary. The project leader should be the champion of termination. Don't protect a project that should be killed off.

- **Purpose**

  The purpose is to reach a decision on terminating a project, or at least initiating a major overhaul.

- **Approach**

  Start by building up the project from an overall perspective. Emphasize what has changed since the original purpose of the project was approved. This will help to ensure that no one places the blame for the termination. Then point to the effects of termination. Don't focus on the sunk costs. They are gone. Press for approval of the termination actions.

- **Suggestions and hints**

  Always consider termination as an option for any project. If nothing else, it forces you to validate the need for and benefit of the project.

- **Measuring your performance**

  Did you achieve the right outcome? Was termination successful without any placing of blame?

POTENTIAL PROBLEMS OF A PRESENTATION

Here are some problems that you may run into and what you might do in response.

- **You are not given a chance to present.**

  This is often due to scheduling or to other logistics problems. Don't personalize it. Try to get temporary approval so that the project can continue. Then present it next time. It is often better not to present than to be forced to present with insufficient time remaining.

- **The key managers did not attend the meeting.**

  If you know in advance of the meeting that this will happen, consider removing yourself from the agenda. If the audience is not right, then you may waste the impact of the presentation. On the other hand, if an enemy of the project is going to be out of town, you might want to move up the presentation.

  If you show up and the key managers do not, make the presentation. After the presentation, go to each manager's office and offer to provide an informal, short presentation. This is your chance for follow-up.

- **You failed to respond adequately to questions and comments.**

  When people ask questions, listen carefully. Let them fully explain their ideas. Don't interrupt. Break up the points they raise in a sequence of numbers. Then address each directly.

EXAMPLES

Modern Examples

In the manufacturing firm and the consumer products firm, the project managers had to cope with authority while the project was centralized and driven from the top. Academically, this might appear to pose no problems due to the existence of top management support. However, the projects were hampered by the lack of support from the divisions and groups.

In each case, the project managers succeeded with management only by showing the benefits of the projects to their own organizations. The lesson learned is that you must appeal to self-interest of management to gain their understanding, involvement, and support.

GUIDELINES

- **Positive management exposure leads to success in getting resources.**

  Only in theory do projects compete only for resources. In the real world they also compete for management attention. Work on getting management attention in order to get resources.

- **Telling management too early that a project is a success can lead to its failure.**

  Treat success as expected. Always caution that more milestones remain to be achieved. Exude a feeling and impression of cautious optimism. If people think that the project is a success, they let down their guard. When a problem later arises, it is a surprise and you find reduced support.

- **Align a project to management's self interest.**

Always show how the project supports the self-interest of the organization and the managers. To do this, indicate the benefits that will accrue. Prepare tables of benefits and show how the project is aligned with the needs of management and of the organization.

- **Look at the worst that can happen in a project and then minimize its likelihood.**

What is the worse that can happen in the project? Make a list of the top five items. These can range from lost resources to the disappearance of management support. Think about how you would respond if any of these occurred. This is not just contingency planning; it is also a way for you to practice problem-solving before you are in the heat of battle.

- **Keep the managers of team members informed of the status of the project.**

Try a dry run of all of your presentations on the project team. This will solicit the input of team members, inform them of the "party line" on the project, and allow them to help in marketing the project.

- **Keep project information you present to a minimum.**

Be very careful about the level of detail and amount of information you present. Do not present detail unless it addresses a specific issue. Avoid cumulative project statistics except for budget vs. actual and plan vs. schedule in general. Concentrate on short-term issues and project status in routine meetings.

- **Always plan to present in a period of five to ten minutes.**

In a project presentation if you are told you have 20 minutes, assume that you will actually have 5-10 minutes to present the project. This will force you in planning to give attention to issues and decisions you need to have the audience consider. If you plan for 20 minutes, you may find yourself filling time with status information.

- **Summarize project status in one page.**

  Lengthy summaries will not be read. The purpose of giving project status is to inform and allow for understanding or decisions, or both. A single page containing schedule (summary GANTT chart), project spending and budget (a graph of budget vs. actual), issues, and key events will often suffice.

- **Include alternatives and implementation in a discussion of issues.**

  At the end of many advertisements is often some action for the consumer to perform, such as a making a telephone call. It is the same in project management. If you discuss an issue in a meeting without getting into alternatives or actions, it seems theoretical. People are not pressed to take action.

- **Be self-assured in presenting issues.**

  A project's external appearance reveals a great deal about the past and present issues within a project.

- **Maintain a low profile**

  Lack of project visibility does not equate to project insignificance. A low profile has advantages, as discussed earlier in the chapter.

- **Read between the lines and determine what is going on through impressions, appearances, and symptoms.**

  Do not wait for a neon sign with a message from management. Figure out what is needed and initiate action.

- **Give team members due credit.**

  A person who takes all of the credit is eventually a one-person project team. If you take too much credit, managers will become doubtful of other things you say or write. The team effectiveness will diminish.

- **If you want to use humor, make jokes about yourself or your own project.**

  This shows that you have the ability to laugh at yourself. Making too many jokes at the expense of other projects can make your project a laughing stock.

- **Portray yourself as confident and knowledgeable about the project.**

  A leader's demeanor tells the management team far more than many project charts and graphs. If a leader appears beaten down or nervous, this will have a negative impact on managers. Also avoid saying "I don't know; I will follow up on that."

- **Keep content as a priority over presentation.**

  Concentrating on the format and style of the presentation while neglecting the content will lead to trouble. People who see the presentation may immediately become suspicious about the project. If problems exist, keep them out in the open. Note also that if you make a slick presentation with no problems, the same level of presentation will be expected in the future. Don't present and do work that you would not do on a regular basis.

- **Be ready to answer questions during a presentation—don't wait until the end.**

  Encourage questions at the start. When asked a question, repeat it in your own words to the audience. Then answer it. If you do not have an answer, write it down in front of the audience and get back to the audience and questioner.

- **Try to be placed in the middle of the agenda.**

  Being first on an agenda is not always an advantage. Yes, you get the audience when they are fresh. You also know you will have your assigned time. However, being in the middle or last has more significant advantages. You can compress your presentation. The audience will be more likely to remember your presentation, since it came later. However, if you are at the end, you do risk being bumped.

- **Remember your goal—to get approval.**

  People try to make a management presentation at a very high level. They want to make it appear strategic. But in fact, management sees many presentations, most of them probably poor. Many managers would rather be doing something else. So get to the point. Take a marketing-oriented approach. Your goal is to obtain approval, not applause.

- **Always be prepared to present the status of the project or discuss an issue with management.**

  Some people attempt to isolate project management from other duties, but this is impossible. At any time you may run into an upper level manager who asks about the status of the project or a specific issue. If you say that you will look into it, you convey the impression that you are not on top of the issue or project.

- **Always end with the default actions that will be taken.**

  You make a presentation and the managers say they will think about it. This is basically your fault because you did not end the meeting with closure. You left too many loose ends. Waiting for management approval is a common excuse for inaction. Instead, offer actions to be taken as you wrap up your presentation.

- **Debrief the project team after a presentation—or someone else will.**

  After a project presentation, hold an informal meeting with the project team. If you do not, word will filter back from other participants. Then you may spend much more time correcting false impressions. Schedule a project team meeting within 15 minutes after you get back from the first meeting.

STATUS CHECK

- How do you prepare for presentations now? How do you prepare documents? What preliminary reviews do you receive?
- Do you make the same mistakes repeatedly in presentations?

- Do you take the time to evaluate yourself after a presentation?
- Do you combine marketing and sales with providing information for understanding of the project?

ACTION ITEMS

1. Identify three or four key managers who are critical to the project in terms of their approval of major milestones. Develop a plan for how you could contact them informally at the start of the day.

2. Prepare yourself for the informal management contacts by making sure that you are aware of status and issues. Begin to update one of the managers. After doing this once a week for several weeks, expand your contacts to another manager. Every few weeks, add another manager.

3. Review your last formal presentation. What preparation in terms of contacts did you make? What efforts did you make to practice the presentation? What surprises arose during the presentation? How did the audience participate in the presentation?

CD-ROM ITEMS

08-01   Powerpoint 95 Issues Presentation

08-02   Powerpoint 95 Multiple Projects Presentation

CHAPTER 9

TRACKING AND ANALYZING PROJECT RESULTS AND STATUS

CONTENTS

9

TRACKING AND ANALYZING
PROJECT RESULTS AND STATUS

INTRODUCTION

In project management training, little attention is given to what project managers spend more than 90 percent of their time doing—day-to-day routine project management tasks. This is a neglected area that has contributed to the downfall of many project leaders. Coping with issues and crises may seem to be separate from the tasks of daily routine management, but these are actually inseparable. The managers who are often best able to cope with crises are those who have their finger on the pulse of the project and know the work.

Let's consider the modern examples. The manufacturing project leader initially was caught up in project reporting and in tracking planned vs. actual costs. However, the project gradually slipped away from the manager's direction. After seeing the decline in results, the manager changed the mode of operation to focus more on results and work and less on administrative work. This shift in focus saved the project from failure.

The construction company manager was an experienced project manager and knew the dangers of being caught in the reporting trap. He delegated the individual project reporting to the project leaders in each country. In that way, he minimized the time spent in mechanical tasks.

Management at the consumer products firm wanted to manage the project in detail with frequent reports. Most of the work time of the project leader was consumed in administration. One project leader approached management with the proposal to provide status and issue reports only once a week, so that he could spend time doing the work. The managers were receptive to this idea and work on the project flourished.

PURPOSE AND SCOPE

The purpose of this chapter is to help you be more effective and efficient as a project manager in the day-to-day work of directing the project. Good daily habits and an efficient work pattern are not intuitive; most managers have to learn them. The goal here is to teach you how to set up work patterns that will help you stay on top of the project. You should be able to administrate the project and still have sufficient time to address issues and opportunities and to communicate with management and staff. You will learn ways to be proactive, rather than merely reacting to events.

The scope of this chapter covers the day-to-day project management activities that you face. Specifically excluded are addressing issues, dealing with crises, and measuring the progress of the project. These topics are so significant that separate chapters are devoted to each one.

APPROACH

Here is a list of tasks on which you will spend your time as a project manager.

Group I—Administrative Tasks

- Determining the status of the work
- Tracking the progress of the work
- Updating and maintaining the project plan and budget
- Carrying out administrative tasks (e.g., performance reviews, hiring, terminating)

Group II—Project Work Tasks

- Doing actual work in the project
- Motivating the staff
- Analyzing the project
- Meeting with and reporting to management
- Evaluating the quality of the work and milestones

The tasks in Group I concern overhead and administrative work. The activities in Group II tend to be proactive—you plunge in and do the specific tasks or take on the issue. Spend most of your time on the tasks in Group II, as these move the project ahead. In contrast, reactive tasks in managing a project occur when you fail to track the project adequately. Then a problem occurs and you must react. By the time you have addressed that problem, another surfaces. You are always behind.

The more proactive work you do, the more you tend to be aware of what is going on in the project. The more reactive work you do, the less you are in control of the project and the more events control you. Assume some manager greeted you in the hallway and asked, "How is the project going?" Could you respond with detailed information? If not, concentrate more on the tasks in Group II to work towards better control of the project.

Note that project control, as people commonly define it, includes some items from both groups.

To track where you are spending your time, at the end of each day write down the rough percentage of time you spent in both groups. After you have done this for several weeks, determine whether any patterns exist based on the day of the week or the time of the month. Work toward increasing your time in Group II activities.

PERFORMING ADMINISTRATIVE TASKS EFFICIENTLY

Determining the Status of the Work

The basic objectives are to know what is going on in the project and to know where the project is going. Tracking projects allows you to resolve issues early and to take advantage of opportunities. A project can be tracked from different perspectives:

- **Project team** The levels of morale and work satisfaction color the perspectives of team members.

- **End user of the project** This perspective looks at the products that result from the project.

- **General management** The priorities are costs and schedules.

- **Line management** Line management is concerned with the use of their resources in your project.

In collaborative project management, the project team members track progress and alert the project manager of any problems or issues arising in their tasks. This is much more effective than the traditional approach, in which the project manager was left to ferret out the problems and status alone. To ascertain status, go to each team member and ask how the work is going. This will yield status information. If you ask directly for a status report, you may get a rosy, unrealistic view as the team member tells you what he or she thinks you want to hear.

Tracking the Progress of the Work

The first goal in tracking is to understand what is going on. Once you understand a situation, you can think about decisions and actions. When you are trying to understand what is going on, you often stumble upon targets of opportunity. This occurs when a team member mentions an idea for a small change or improvement. This often costs nothing and does not involve any high-level management action. When this occurs, take action and implement it right away.

Here are several approaches to tracking work:

- **Approach 1: Track all work and tasks in the project with the same level of detail and effort.** This appears fair and makes some sense when you first consider it. However, it is not a very intelligent use of your time given that it does not take into account the stages of various issues or the timing and duration of tasks.

- **Approach 2: Track work based on the mathematical critical path.** This is a traditional approach focusing mainly on the critical path tasks. These are tasks that happen mathematically to fall on the longest path. The problem with this approach is that it is not sensitive to risk and uncertainty. Also, it is not sensitive to time. Logically, you should spend more time on tasks that are in the near future.

- **Approach 3: Track work based on the managerial critical path.** The managerial critical path is a set of paths in the project plan that contains tasks that have substantial risk. How do you know which tasks these are? Look at the existence of issues and problems associated with tasks, as well as the uncertainty inherent in the tasks. Using this approach, start with the tasks that are happening now that have risk and expand to those that have risk in the next two months. Then extend your examination to all of the tasks on the mathematical critical path that carry risk.

The third approach is favored since it is most reasonable in terms of the resources available and it is centered on minimizing risk.

How should you track routine tasks? In collaborative scheduling, you want the team members to participate and let you know what is going on one-on-one. Also, randomly check on some routine tasks.

The first step in tracking work is to collect the following information:

- **Collecting information from team members.** You want to track what is going on with each team member, individually. Drop by each team member's office. Why not have them come to your office? You don't interfere with their work as much by visiting them. Also, when you visit others, you can see what they are working on. This is useful in tracking people who are part of the project's core team.

  During the visit, encourage team members to talk about their tasks in their own words. Take notes yourself and don't use forms. Don't use checklists or task lists and go down each list. These approaches are all too formal and may lead to answers of "okay" or "so-so" that aren't helpful. Prepare for these meetings by reviewing the tasks and issues each team member is working on. After the meeting, return to your office and prepare the team member's part of the status report.

- **Collecting information by observation.** Observe what is going on in the project firsthand. This is very important, since you can use this to update the tasks that are active in the project. Observation does not help in updating the estimates for future tasks, but it can tell you what is going on now.

- **Collecting information from meetings.** You attend many meetings on the project and with management. At the end of each meeting, ask if there was anything discussed in the meeting that has bearing on your schedule. Did you learn anything that will have an impact on resources? Do you foresee changes in methods and tools?

When should work be tracked? Today, the response is "continuously." In the past, if status or progress reports were due monthly, people would collect data on status and do tracking just before the end of the month. After the presentation of status, tracking would become a low priority until the end of the next month.

A better approach is to be aware of what is going on all of the time, since situations in projects can change rapidly. Continuous tracking will allow you to understand what is going on at all times.

Along with continuous tracking, update your schedule and plan as you get the information, at least twice a week. A daily update is not necessary, as this requires too much effort for what it gives you.

Here are four problems that can arise during the course of a project that relate to tracking.

1. Management wants frequent and detailed reports.

   This may occur because of problems within the project. It can also be due to a faulty project reporting and control process. Project control should be sensitive to the size, type, risk, and importance of a project. It is also dependent on the stage or phase of the project. For example, management would want more information during critical stages of construction.

   How do you cope with this? Start with increasing informal contacts to let management know what is going on. This will provide management with more information. Next, propose a summary reporting process that can replace the existing detailed reports. Prepare the information yourself. Don't shift the burden to others on the team, since this can lower productivity and morale.

2. You inherit a poorly run project.

   It has been assumed until now that you have been the project leader from the beginning. What if you are taking over a project in trouble? What should you do first? The reason for discussing this topic now is that your initial actions upon takeover are going to lie in tracking. Here is a sequence of steps in project takeover:

Step 1: Determine the status of the project and assess the team. Don't ask for more resources or money yet. Find out what is going on in the project. In finding out the status, you can assess the project team members in terms of their productivity.

Step 2: Conduct analysis of the project to determine what could be done to improve project performance and results. Follow the steps for analysis given in this chapter.

Step 3: Assess the current open issues and see if you can make some quick progress on some of these.

Step 4: Develop a new project plan and approach and present it informally to management. Indicate what can be done with limited incremental resources.

Step 5: Implement changes as soon as approval is given. Implement a more formal project reporting process. Management is giving you more resources. Reciprocate by giving management additional information on their investment.

3. You have to deal with false information.

False information on the project can originate inside or outside the project. The damage it causes can be extensive. Your time may be consumed dealing with problems that are perceived but not real. This reduces both productivity and morale. Management may get the wrong impression about the project and institute countermeasures without asking you. This can be a major issue in that a new layer of management may be created.

How do you head this off? First, keep your ear to the ground and determine what is being said about the project. Second, draw in the project team to do the same and to support you. Third, stay in informal contact with managers you can trust to get early warning signs of problems.

4. You lack technical knowledge.

You are managing a project in which you lack technical knowledge. What do you do? You will obviously make an attempt at trying to learn the technical words and concepts. However, you will still lack detailed hands-on experience. Identify several informal technical advisors who can help you in reviewing plans, assessing milestones, and determining how methods and tools are being employed.

Updating and Maintaining the Project Plan and Budget

Let's assume that you are using a standard project management software package. Here are the steps in updating the project plan and budget:

Step 1: Update the current schedule bottom up. That is, go to the most detailed level of tasks and update these. Mark the relevant tasks that are complete; change the duration, resources, dependencies, and dates where necessary. If you are doing collaborative scheduling in which the team accesses the schedule, then set a deadline for the update. Review the update.

Step 2: Move into the future and enter new task detail in the schedule. You may actually change the structure of the schedule based on your knowledge now.

Step 3: After making changes to the detail, set the actual schedule. This will recompute the critical path and overall schedule.

Step 4: Go to the databases that you have established for issues and action items. Update these as well, based on status.

At this point you are prepared to analyze the schedule. Note that the approach we have suggested focuses on doing routine work and no analysis. Give attention to the detailed tasks in updating. Keep analysis separate from updating.

Carrying Out Administrative Tasks

A project manager performs many mundane but significant tasks. These include performance evaluations for team members, recruiting and interviewing potential new team members, dealing with personnel problems on the team, checking up on the timekeeping and human resources of the team, and monitoring vacation time and sick time. Administrative duties also include maintaining project files, determining training needs, and reviewing what other projects are doing. These are important duties. However, control the amount of time spent on them. Plan when you will do administrative work and group these duties into one portion of the workday. For example, try devoting the early morning hours several days a week and see if this is enough time to handle the work involved.

PERFORMING PROJECT WORK TASKS EFFECTIVELY

Doing Actual Work on the Project

It was once thought that when you became the project manager, you were removed from the actual work on the project. In cases where specific skills are required, or danger is involved, this is still true. However, hands-on work on the project is one of the best ways to see what is going on and to judge the impact of issues and opportunities on the project.

What tasks can you take on? You won't be able to assume a standard routine manufacturing or distribution job, for example, due to the demands of project leadership. As you choose tasks for yourself, keep in mind two priorities:

1. Assign yourself some routine tasks.

   These could involve documentation, testing, preparation, or analysis.

2. Look for tasks that you can perform with other people.

   If you work alone, you still assist the project, but you are not in touch with the team. Work with the staff on the project one-on-one on a rotating basis. Do the job with the team members. Have them train you on how to do the work, if this is needed.

Your work on tasks may seem to slow things down at first, but the benefits include the following:

- The project benefits from additional resources working on the tasks.
- You gain an awareness of the work and potential ideas that could improve the work and results within the project.
- You maintain open communication with the project team by working with them.
- You improve your ability to assess the effect of specific issues, opportunities, and crises on the project.
- Assessment of the status of the work is easier, faster, and more accurate.

How much time should you spend doing actual work? Obviously, it depends on the project. In general, 20 to 25 percent is a reasonable amount. If it is much higher, you will be too involved in project work and will neglect project management. If it is much lower, then you will waste time setting up and shutting down for work.

How should effort be split between the direct work on the project and working with others on the team? A split of about 30 percent individual work and 70 percent working with others is reasonable. This provides you with more project team contact. When you work with others on the team, don't assert authority as a project manager. Let them lead you through the task and you follow.

For resources that are shared among projects, work with other project leaders on a weekly basis to set priorities for each person during the coming week. This will ensure that each person obtains a consistent story from all project leaders.

Motivating the Staff

Some project leaders either downplay this or pay it lip service. They assume that getting a paycheck is sufficient motivation. Or, they may get people together and give a motivational talk.

Here are some better ideas on motivation:

- When gathering status on the project from team members, show not only a sincere interest in their work, but also try to see what you can do to help them.
- Group motivation should be done through addressing issues and paying the team overall compliments. Group motivation

has its drawbacks. First, if you do it often, it loses impact. Second, diligent workers may feel slighted when you compliment a group of people in which some members slacked off.

- Follow up on suggestions by the team members concerning improvements or problems. Give them credit for their suggestions.

- Compliment people who raise many problems—the more the better. Hiding problems or not taking action is a recipe for disaster.

Analyzing the Project

Divide the analysis into steps. A guideline here is to divide the update and the analysis. Do the update and the analysis at different times, because they require different skills.

Step 1: Validate the schedule

Review the milestones, dates, and summary tasks of the schedule to see if they make sense. If events suddenly shift, check to see if you left out a dependency or missed some tasks. Use the project management software to determine completeness of the tasks and the impact of change.

Step 2: Assess the mathematical and managerial critical paths

You want to determine how these paths have changed. If a task has become critical, why did this happen? If the path is longer due to greater detailed task durations, this indicates slippage in the work. Is the path length due to more information on the work that increased the number of detailed tasks? This is a common occurrence and not unexpected.

You might lower or raise the risk when you update the task, depending on the issues. Lower the risk for the task if the issue has abated or been addressed. Raise the risk if the task has grown in importance, if there are new issues and if the assumptions made about the task are no longer valid.

Step 3: Compare the planned and actual schedules

Use tables and GANTT charts that compare the actual and planned schedules. To analyze them, go back to the start of the project and work forward in time. Consider where the schedule first began to slip. Then move ahead to note areas where slippage increases. Start with a high level outline form of the project and then step down into more detail. Figure 9.1 gives an actual vs. planned GANTT chart.

Step 4: Do actual vs. planned cost analysis

Create a spreadsheet using project management software. You can produce a table in most project management software packages that gives the planned or actual work performed by resource (rows) over time (columns). This table can be exported to the spreadsheet. With both the planned and actual work exported, you can compare the results in terms of hours.

To do cost analysis, use the spreadsheet to convert work into money considering regular pay, overtime, and other cost factors. Figure 9.2 gives a cumulative actual vs. budget analysis. Note that in this example the actual cumulative expense lagged behind the planned expense for some time. At the current time this project is now over budget in terms of cumulative costs.

Step 5: Analyze variations

If you have found variations, why did they occur? The obvious reason is that the task dates and durations slipped. A second reason is that more information is reflected in changes in schedules and dependencies. Third, you may have added more tasks and detail that can impact the overall schedule through a rollup to summary tasks. Fourth, you may have restructured the tasks in the schedule.

Step 6: Perform a "What if . . . ?" analysis

At the end of the actual analysis, analyze the impact of shifting resources, adding resources, changing project structure, or deleting resources. See what happens if you shake things up. This can lead you to some interesting ideas for project change.

Figure 9.1
Actual vs. Planned GANTT Chart

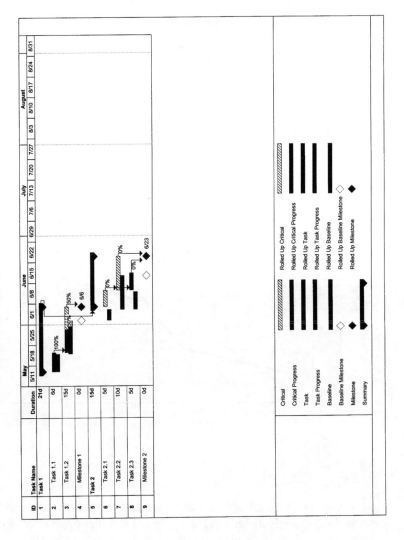

Figure 9.2: Cumulative Budget vs. Actual Expenses

In this diagram the actual cumulative expenses are shown in the dotted line. As you can see the project started more slowly than was budgeted. This is not unusual. Then it exceeded the budget amount, probably in an effort to acceleterate the project. Finally, spending tapered off.

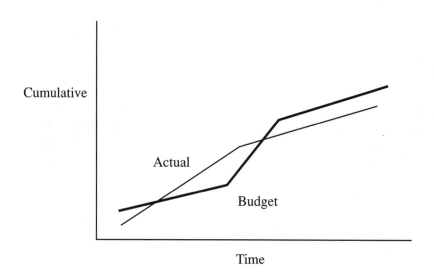

Perform Financial Analysis

You will want to review the actual expenditures by area. Set aside time to do this analysis on a regular basis. Take the reports you are provided by accounting and review these in light of the plan. Here are questions you should ask:

- Have part-time people working on the project been charging too many hours to the project? Relate the part-time people to the tasks that they are performing to formulate your answer.

- Are the facilities use and equipment charges valid? Go back to purchase orders and the plan for your information.

- Is the overhead and burden assigned to the project excessive?

If possible, maintain your own spreadsheet for budget items. If you hold yourself accountable to maintain numbers, you are more likely to have the discipline for the analysis of the data as well.

Common Analysis Problems

The above analysis steps are inductive in that they proceed from validation to consideration of variations. Analysis can also be inductive when your analysis is based on an issue.

- Problem 1: The schedule is standing still, but you know that the work and progress are going well. What is going on? Probably tasks are missing from the schedule. When these are added to the schedule, the schedule will likely show slippage. Yet, when you post actual results, this should be fixed.

- Problem 2: The work is showing limited progress, but the schedule shows even less progress. The schedule probably does not reflect dependencies or work correctly.

- Problem 3: The number of issues is growing. Not all the issues are getting addressed. However, the schedule is remaining unchanged. You have not reflected the issues in the dates of future tasks. Allow for these additional issues and recognize that, when these are factored in, the future tasks will slip, causing the overall schedule to slip.

Reporting to Management

A major guideline is to meet often informally with management to keep them up-to-date and to garner support. These meetings should be short so that the overall percentage of time spent with management is actually small. Most of the contact should be in informal meetings, rather than in formal presentations. More formal presentations indicate either that the project is becoming more exciting or the project is in more trouble.

Effective Status Reports

Here are three ways to provide status to management:

- Status report based on work and progress. This tends to be a reporting method that focuses on accomplishments. The detailed work and milestones accomplished are enumerated.

- Status report based on issues. This report identifies the various issues still open and their priorities. It also highlights the issues that were resolved.

- Combined status report based on both work and issues.

To ensure consistency across many projects, adopt a standard reporting method, preferably the combined approach.

You will want to provide both quantitative and qualitative information to management at the same time. The form in Figure 9.3 provides a reasonable summary sheet for a general project.

The sections of the report are as follows:

- Title, identification, and purpose. This provides some basic information to identify the project.

- Summary GANTT chart. This is a summary high level GANTT chart that provides status.

- Cumulative budget vs. actual chart. See Figure 9.2.

- Summary of issues. This area provides highlights of active issues.

- Milestones and accomplishments. This section addresses achievements.

- Anticipated activity in the next period. This section describes what is likely to occur in the next period.

Evaluating the Quality of the Work and Milestones

What is quality? What is acceptable quality? To answer these questions will require effort and definition for each milestone. Checking out the quality of dozens of milestones is not practical, requiring time and resources. Here are some alternatives for evaluating milestones.

- Level 0. Do no review.

- Level 1. Determine if there is evidence of a milestone. This is verification of presence, not content.

Figure 9.3: Example of Summary Sheet for a Project

Project Summary Sheet

Project Name: _____ Date: _____

Project Manager: _____

Purpose of Project: _____

_____

Summary GANTT Chart	Cumulative Budget vs. Actual

Results/Milestones Achieved in Last Period: _____

_____

Critical Issues: _____

_____

Anticipated Results/Milestones: _____

_____

- Level 2: Perform a quick check by evaluating only a few items related to the milestone or work.

- Level 3: Conduct a full milestone or work review.

When work is routine and you trust the people involved, impose level 0. Many key milestones will deserve at least a level 2. Few will justify a level 3 review. You can escalate a review from one level to a higher level if you sense problems.

What is involved in a full-scale review? Three setup tasks begin the process:

1. Determine who has knowledge of the project, technology, and situation to be involved in the review. Involving people in a review means time lost from working on your project or other projects, so enter into the review process judiciously.

2. Set the time for the review and try to manage it in such a way that it causes a minimum of disruption to the project.

3. Define the scope of the review. What will be included and excluded? For example, will only quality be reviewed or will the review include the way that the methods and tools are employed?

With the setup tasks accomplished, move to the review itself. Make a checklist of items:

- Materials that will be supplied before the review

- Materials and documents during the review

- People, equipment, and facilities access required for the review

Divide the review into two parts. The first part consists of an overall assessment by the review team with feedback as to the level of detail for the review and the areas of the project milestone or work for review. The second part consists of the actual review. The review report is addressed in more detail in Chapter 13.

After the review of a milestone or work, implement the results as soon as possible. This may mean changing the project plan. It may mean resource shifting or change. It can also be a time of getting charged up again in the project. If the results of the review are favorable, the team's efforts are reinforced. If the results are recommended changes, you can tout success based on the changes and lessons learned.

EXAMPLES

The Railroad Example

Railroads in the 1800s faced a major problem: How could trains travel toward each other on the same set of tracks without a head-on collision? Double sets of tracks were not affordable. To prevent problems, stations had to be informed of train locations at specific times. The solution to how to communicate between stations was the telegraph. During the Civil War the telegraph proved useful in providing information on enemy troop movements. It was also used in conjunction with the railroad. After the Civil War, the telegraph was immediately employed throughout the railroad industry.

The telegraph was a boon to project management and the direction of the work. Requests for additional resources and supplies, confirmations, issues, opportunities, and legal agreements on right-of-way were all handled through the telegraph. It was a perfect medium at the time since it carried no additional cost for its use for these purposes; it had already been justified as an expense necessary for the scheduling of trains. The telegraph had another impact on railroads. At the early stages of construction before telegraph cable had been laid, remote managers and supervisors had more autonomy. After the telegraph was installed, central control and direction increased. Given the speed of the telegraph, information could be transmitted and rumors could be checked out and verified quickly.

Project administration was centered on the division and headquarters offices. With the telegraph and mail, information was provided from the field and processed in these headquarters and regional offices. This was of benefit to the field managers, since their administrative overhead was lessened.

Modern Examples

The project leader in the manufacturing firm quickly saw that the project could not be tracked by one person. It was spread out over too large an area. Moreover, the project leader lacked a detailed network technology background. Managers for regions of several countries were identified as subproject leaders. Even with this change, it was necessary to spend time interpreting the information provided.

The construction firm had to replace the scheduling and project management process. The firm wanted to move away from professional project schedulers and evolve into a more collaborative planning process. How could this change be implemented in the current organization culture? Two projects were selected as models. The project tracking approach was implemented for these two projects. The results were then applied to two additional projects. After this refinement, a management review and proposal extended the process to all projects.

GUIDELINES

- **Consider what omens mean and what pattern they imply.**

  As in real life, omens can appear for projects. These can be problems or successes with methods and tools, people, or other resources, for example. Rather than just tactically responding to these, look for a pattern so that you can be proactive in dealing with problems.

- **Take stock of the project when you are too tired to do other kinds of work.**

  When you are tired, you may not want to deal with the project anymore. This is a good time to sit back and review the project overall.

- **Allow ample time to do project analysis before a meeting.**

  Last-minute work on a project before a meeting can result in more chaos than benefit. Use the time right before a meeting to review what you have done before and get focused on issues. Project analysis time should be open-ended and not subject to pressure.

- **Distribute project knowledge throughout the project team.**

  People sometimes associate control of a project with knowledge about the project. For this reason, some project managers attempt to keep knowledge to themselves, believing this will allow them more control. In a modern collaborative environment, this is clearly out of place. Control and knowledge are related, but project knowledge should be distributed throughout the project team.

- **To gain wisdom, sit back and look at a project.**

  A project manager who is constantly working on detailed administrative or issue work loses the benefits of considering the overall picture. Gaining perspective and an understanding of what is going on in the big picture are two valuable aspects of taking time to take a step back.

- **Do project analysis yourself.**

  Do not depend on others to do your work in analysis. First, when this person is not available, you are helpless. Second, in meetings you will be unable to respond to questions related to the analysis. Unless you do it yourself, you are remote.

- **Suspect trouble and check up on the situation if people do not inform you of the progress of particular project tasks.**

  People who are achieving results are usually happy with what they are doing. They are likely to relate their success to you. On the other hand, if you hear nothing, you have a right to suspect trouble.

- **Retain project history so that you are not doomed to repeat mistakes.**

  If a project goes on for a year or more, similar issues and questions will crop up numerous times. Knowing how previous problems were addressed will help you now. Also, note that some issues will recur in different clothes. This may occur with someone who lost out on an issue, for example.

- **Make use of statistical analysis of project data.**

  Project statistics usually abound and are there for the taking. These statistics are often boring to work with and overlooked, but they are most useful. If you do not analyze the data, you will wake up to unpleasant trends too late. With the availability of more statistical tools in spreadsheets and other software, statistics are getting easier to work with.

- **Track and compare multiple schedules for a set of tasks.**

  In many settings, several projects compete for the same people, supplies, facilities, or equipment. You also may want to learn from previous, similar projects. Yet if the schedules and resources are not compatible, it will be impossible to make any meaningful comparison. What a term or task means to one person may mean something different to someone else. Work out a system which allows for comparison.

- **Look for the real project bottlenecks.**

  You have been taught that the project bottlenecks can be found on the critical path. It seems logical that if you can shorten or rearrange the work, the bottleneck will disappear. The reality isn't this simple. You are watching the critical path, but problems arise from a task off the critical path. This occurs because the critical path does not include risk and uncertainty; it includes only length and duration. Real project bottlenecks cannot be detected easily by looking for the red line on the GANTT chart. How can you prevent bottlenecks? Go through the schedule and label tasks according to risk. Then filter or flag these tasks. Consider how close these are to the critical path. Continue to keep a close watch on those with the least slack.

- **Allow for a difference between resource allocation and resource usage.**

  Resource allocation is the assignment of resources. Resource usage is the consumption of resources according to specific schedules and calendars. These may not match, for several reasons. The resources are allocated at the level of higher level tasks but are consumed at the level of more detailed tasks. Also, the overhead associated with a resource is often not factored into resource allocation. For example, you may allocate someone for six months to a project. However, the team member is on vacation for two weeks and in training for another two weeks. Allocation and consumption would be different.

- **Add tasks related to error fixing and rework at a detailed level.**

  The inability to cope with the need to rework in a project affects likelihood of eventual success. Many people just extend a task duration to reflect fixing errors or rework. This does not convey what is happening and creates communications problems. Instead, add in the actual tasks that have extended the duration of the task. If you allow the task to be slipped, you lose history and accountability later.

- **Examine project boundaries periodically.**

  During the duration of a project, the nature and boundaries of a project can change. As a project leader, you should examine the project boundaries as part of your "What if . . .?" analysis.

- **Manage a project for long-term payoff.**

  Patterns of work behavior, relationships between people, and experience with methods and tools often long outlive the original project. Side effects of a project may long outlive the project impact. Keep the long-term view in mind throughout the life of the project.

- **When you see a problem coming, give warning.**

  Do missed deadlines have penalties? For example, you deliver a milestone a month late. In many cases, the project grinds on. Maybe no one will say anything. But you lose credibility when you don't see a problem coming and take some ameliorative action.

- **Determine deadlines by need rather than playing games with deadlines.**

  A middle level manager wants to look good. This manager imposes unrealistic deadlines on the project. If the project team can make it, this will reflect favorably on the manager. This sounds fine, but this will only work once or twice, if at all. People become wise to this strategy and start to give dates more conservatively to compensate.

- **Manage tasks that have risk as well as those that are easier to handle.**

  Balance your time between different project management tasks. If you devote too much of your time to tasks that are comfortable, you will not be coping with the tasks that have risk in the project.

- **Assign as many cheap solutions and resources as possible.**

  Depend as much as possible on simple, cheap resources. You will be pleasantly surprised to see the benefits from the expenditure of a small sum.

STATUS CHECK

- To what extent are you on top of the project? Are you aware of what is going on in the project today? If you had to walk over to the project team and go to the most critical area of the project, where would you go?

- Where do you spend your time? How much time is spent in the interactive, more productive work in the project? Have you attempted to spend more time here? What is preventing you from spending more time in these tasks?

- Have you adequately delegated the tracking of the project and work to people on the team? Have you adopted a more collaborative tracking approach?

- What milestones and work have you recently reviewed? Was the right information available? Were the correct people involved in the review? What results and actions flowed from the results of the review?

- Do you find it easy to relate the issues and action items in the project with the project schedule? Have you identified the areas of risk in the schedule?

ACTION ITEMS

1. Using the list of Group I and Group II activities in this chapter, list how much of an average week is spent in each activity.

Make a list for several project managers around you, also.

2. Evaluate your technique for assessing milestones in your project or the process in a project with which you are familiar. Should you adopt a more formal process in evaluating milestones and work?

3. Review your update and analysis process for your project. How well is it organized? Have you divided updating activities from analysis? Does the analysis that you do get translated into actions and schedule changes?

CD-ROM ITEMS

09-01	Project Effectiveness Evaluation
09-02	Project Review Checklist
09-03	Project Review

CHAPTER 10

COLLABORATIVE SCHEDULING
AND WORK

CONTENTS

10

COLLABORATIVE SCHEDULING AND WORK

INTRODUCTION

A major challenge facing organizations is how to manage projects involving multiple organizations, both internally and externally. This is often made more complex if the participants are distributed around a region of the world and belong to different companies with different cultures, goals and objectives, and technology. Traditional project management was not designed for this situation. In fact, until the 1990s there were only a limited number of successful distributed projects. Most projects were managed as centralized, traditional projects. Some of the business factors that encourage larger scale projects include the following:

- More companies are attempting to work with each other in supplier-customer relationships.
- There is a desire to implement more uniform systems and technology across medium and large corporation.
- Many projects require skills that reside in different companies and countries.

Some specific examples of collaborative projects are as follows:

- Company expansion into a new state or country
- Implementation of Enterprise Resource Planning (ERP) software
- Joint programs with business partners
- Electronic commerce projects (See Chapter 17 for a discussion of e-commerce.)

These collaborative projects represent a challenge because of factors such as the following:

- The culture and interests differ among team members and companies.

- Many individuals assigned to the project have normal, non-project duties that they cannot give up for the project; dividing their time among project and non-project work is a major challenge.

- Different companies may employ a variety of IT methods and tools which do not easily support integration and lead to incompatibilities.

- Projects have hidden dependencies that are revealed only later in the project at critical times.

- The goals of the project may not be relevant to many of the team members.

At the heart of these issues is the problem that the project leader does not have total authority over members of the project team.

Collaborative Management

What is a collaborative management approach? Here are some key ingredients:

- Each person on the team is responsible for identifying detailed tasks, updating tasks, addressing issues associated with their tasks, and participating in joint project work.

- A substantial percentage of the project work is assigned to more than one person. In some cases, 30 to 40 percent of the tasks are joint among two or three people.

- The project manager shares all project information except the really political elements with the team.

- Project leaders share information amongst themselves. This includes schedules, issues, and lessons learned.

- Project leaders work together and with line managers in assigning people and other resources to tasks on a routine basis (typically weekly).

What are some of the benefits of a collaborative approach?

- People working together and sharing information tend to trust each other more. They grow closer together.
- Working on issues together helps to build skills of the people in the project.
- The project is more likely to end successfully on time and within budget.
- A ready forum is available in which to gather lessons learned.

PURPOSE AND SCOPE

The obvious goal is to achieve the objectives of the project within budget and schedule constraints using a collaborative project management approach. This definition of purpose should be expanded to include the interests of the organizations and individuals participating in the project.

The scope of the project is defined. However, the roles and responsibilities of the project team are more broad than that of the traditional project manager, as seen in the above list of the duties of the project team members.

APPROACH

Action 1: Define the Project Concept and Identify Other Work

The actions in this step are intended to build a common vision of the objectives and scope of the project, identify all of the things that team members are doing, and explore issues in the project identified in the project concept. In addition, you will be defining together how the project will help each person on the team.

Review the Project Objectives and Scope

Go over the objectives and scope of the project with each team member individually. Show how the team member's self-interest is aligned to the objectives of the project. Also show each team member how he or she fits into the project.

After you have met with each person, assemble the project team. To avoid a rehash of what you did with each person, go into the alternative purposes and scope that were considered in the project concept definition. Also, look at the project from several alternative perspectives. These include the following:

- **Business perspective** Show how the project is contributing to the organization. This will help reinforce the feeling that each team member is making a contribution.

- **Technology perspective** Look at the project from the view of the methods and tools that will be employed in the project. Show that the project is employing modern techniques and that these techniques are well established.

- **Management perspective** Explore the management controls and reporting that will be done in the project.

As you are doing this, you can indicate why each person was chosen for the team and each person's role in the project.

In this first action you are also defining the benefits of the project for the team members. In the past, little attention was paid to individual team members and what team members would get out of the project. Yet, this is extremely important when working with team members because your project is competing for their attention with their other work. A prime strategy is to appeal to self-interest. Here are some things to do:

- Have each team member give you a resume when he or she joins the project.

- Have each team member identify career goals and objectives for the next five years.

- Have each team member create a new resume that they would like to have after five years.

- Based on previous collaborative actions, work with each team member to identify tangible things that they will learn and do as well as the knowledge they will obtain from the project.

- Now have each team member create a resume that would represent work at the end of the project.

- Identify issues and barriers to achieving these personal goals.
- Make these issues generic and add them to the list of issues in the project.

Build a Project Plan for Each Team Member

At the team meeting, indicate that you are sensitive to the fact that most, if not all, of the team members also have line responsibilities as well as work on other projects. Visit each person to determine what he or she is working on and what the schedule is. Try to obtain a copy of the project plans for other project work. Give as the reason for this the fact that you must build an overall project plan that reflects the realities of the availability of the project team members.

It is useful to build a small plan for each person with his or her other work. Each task the person performs in a line organization would be one task in this plan. You would also have the person's tasks in other projects in a summary level.

With this done you have a project plan for each team member. When you construct the schedule for the project, you can combine it with these other plans and then filter on each resource to see the total commitment for each person on the team. You will be able to see points in time when people are overcommitted. Then you can plan and negotiate for people's time more effectively.

Negotiate with Line Managers and Other Project Leaders

Work with each manager to define a set of near-term priorities for each key team member. Next, focus on the short term of two to three months. If you negotiate for work beyond that, conditions and situations may change.

Besides accomplishing the setting of priorities with these other managers, this step provides two other major benefits. The first is that you are establishing a collaborative environment for sharing resources with them prior to any crisis or major issue. This helps to build a pattern of successful relationships. Second, you are sharing information with them. You want to build upon this relationship to share schedules and future need information far enough in the future to support planning.

Define Issues Together

The issues are particularly important here because you want to use discussions of these as tools to build a common approach for working on problems and opportunities. Here are some guidelines:

- Do homework on several issues and introduce these to the group.
- Consider as issues the following:
  - —People on the team have other duties and responsibilities. How can they be effective on the project?
  - —The project may be of importance to the organization overall, but it is of marginal interest to some in the team. How will this be addressed?
- As you discuss each issue, summarize how people are to work together.
- Identify how people should report on their project work.
- Identify how people will define their own work.

You can design a template to identify what the upcoming step is in terms of defining the project. Include detailed tasks for each team member and schedule when the team members are to report on these tasks.

Action 2: Develop a Project Plan and Implement a Collaborative Approach

Build or Evaluate a Project Template

Do you have an available project template for this project? If not, define a strawman, candidate template for the team members to review. Recall that the template contains high level milestones and tasks. For each task in the template, identify the team member who will responsible. Also, identify which tasks are going to have joint responsibility. It is useful to have 30 to 40 percent of the tasks jointly assigned to foster teamwork. You should also validate the template by evaluating the issues in the project that surfaced in the previous step. Find the summary task in the template to which each issue corresponds. You can also scan down the tasks to see if you and the team have missed any issues.

Once you have a template, meet with each person on the team and indicate that person's areas of the project. Get each person to think about detailed tasks and relate the issues from the previous step to the tasks that they are responsible for.

Go Through a Simulation of Building the Detailed Plan and Updating the Plan

This is an important part of the project since it basically links the work on issues with the initial meeting on purpose and scope. At this meeting, take one area of the plan and act as a team member in defining the tasks in the template. Progress from defining tasks to completing the baseline schedule for the work. Next, explain how the tasks will be updated by the team member responsible for them.

Construct the Detailed Tasks for the Project for the Next Three Months

Each team member can now define the tasks needed to accomplish work that is to be done in the next three months. It is very important that you have identified tasks that are to be jointly performed by team members. Encourage team members to work together to define these tasks in more detail.

Each template task should be broken down into tasks that are not more than two weeks in duration. If you go over two weeks, the task is too fuzzy. If the task is too short, the effort requiring updating will be too great.

Here are some additional guidelines in defining tasks:

- Each task should be able to be defined as a simple sentence starting with an action verb. An example is "Prepare ground for planting trees." If you find that a task has complex wording such as "Dig up ground, fertilize, and water for trees," then split up this task into three separate tasks: dig up, fertilize, and water.

- Have each team member associate issues with the detailed tasks under the relevant template task for the issue. This further helps to validate the tasks and issues.

- Each team member should identify tasks that have risk or seem risky. This will give rise to additional issues or validate the existing issues.

- Schedule meetings with team members to discuss their joint tasks together with you.

Making task definition a distinct step, apart from schedules, dependencies, and resources, will give a more complete task list and prevents team members from getting distracted by other facets of the project.

Establish Dependencies and Assign Resources

After reviewing the tasks, have the team members put in the minimum number of simple tail to head dependencies. If they are in doubt about a particular dependency, leave it out. They can discuss it later. This may indicate that you are missing a task.

In this action each team member will identify critical resources of any type that are a cost to the project or that the project will have to compete for with other projects and normal work.

In reviewing the work in this action, start with the dependencies. Ask team members why the dependency was created. This will lead to a discussion about the surrounding tasks. The net effect of this is to not only validate the tasks and dependencies, but also to get a better understanding of how the work is to be done. The same is true with assigning resources to tasks.

Define the Duration and Dates Based on Previous Actions and the Previous Step

With the tasks, dependencies, and resources defined, each team member can now estimate the dates and duration's for each task. Give some examples to team members so that they have a better awareness of the approach. Here are some guidelines to help the team members:

- Do not pad the dates for contingencies. Put in realistic estimates.

- If you cannot estimate for a specific task, break up the task until you have isolated the part that you cannot estimate. There is probably an issue here that is the reason an estimate cannot be given.

Review each person's work with him or her when you have received all input from team members. By waiting until you have all inputs, you can see the schedule overall.

It is likely that the schedule will not be realistic. It will stretch too long. Don't attack the group by saying that the schedule is not acceptable. Rather, identify where the specific parts of the schedule are in trouble. Go to the person involved on an individual basis and get at the assumptions behind the estimates. If you are lucky, you will find that assumptions have been made that caused the schedule to be longer but that were not necessary.

After reviewing the schedule, you can set the baseline plan and hold a project team meeting to review it. At this meeting, hand out the schedule along with a list of issues and a mapping between issues and tasks. The purpose of this meeting is for the team members to gain a better understanding of the work as well as to focus on near-term risky tasks as a team.

Action 3: Build Collaborative Teamwork through the Initial Tasks

Work now begins on the project. Circulate a printout of the schedule for the next three months. Have the team members mark tasks that have been completed. Add new tasks that were unanticipated or that apply to the future time horizon. If a task has slipped, have the team members create a new task and link it to the current task, also giving a reason for doing so. By repeating this several times, a person gets used to the process of schedule updating. You can then have a team member do the updating online in the network.

There may be a lack of knowledge of project management software on the part of the team members. Don't wait until the people are trained on the software to begin the collaborative approach. Implement manually with paper to get the process of collaboration going.

This will take more time initially but rushing this learning phase of collaboration will be counterproductive.

Address Initial Issues

Early on, establish a pattern for addressing issues. Identify some sample issues that are relatively minor and non-political. Get people in a group and start analyzing the issues. After some discussion, show the team how decisions are made and actions are taken. You can also indicate how the plan is updated as a result of deciding the issue.

As the project gets underway, ask for the reaction of the team to the process. Have team members share their views and suggestions on making the process better. This is a bottom up approach to implement collaboration and one that has often worked. This approach establishes a pattern for dealing with issues in a friendly and non-hostile setting. You can scale up the issues to address those that are more major.

When you do this with a team, you are accomplishing several goals. First, you are showing the team members that they can solve problems on their own without management. Second, they gain confidence their ability to get things done as a group. Third, you are paving the way for more serious issues to be handled, based upon the pattern of success.

Conduct a Review of Initial Milestones

Review the work of the team members and milestones reached, based on the criteria given Chapter 2. Also, give attention early in the project to tasks that slipped. You are trying to determine a pattern for the slippage. This is done not to punish a team member but to determine now whether estimates for later tasks need to be revised. Try to get team members to the point of feeling comfortable in dealing with milestone reviews of each other's work. Another goal is to position team members to review each other's work, both positively and negatively. Team members will be able to see that people make errors without incurring punishment.

Action 4: Monitor and Manage the Project from a Collaborative View

This step considers some of the most common situations you will face in managing projects collaboratively.

Exit of a Team Member

Prepare the team to address this situation before it arises. At the project kick-off, point out that team members will come and go. Turnover is inevitable. Have the team members identify and discuss issues associated with someone leaving. Some of these are as follows:

- The departing person takes knowledge with them.
- It is difficult to capture all of the knowledge before a person leaves the team.
- The work of the departing team member falls on the shoulders of the people remaining.

The exit of a team member brings not only problems but also benefits and opportunities. First, by the time someone leaves, progress has been made in the project. Work has been started and the person may not be critical to the project anymore. When someone leaves, it gives the team a chance to find a replacement with different skills that are needed for future tasks.

Transitioning between Project Phases

Longer and larger projects are typically divided into phases. Each phase often has a formal ending prior to the start of the next phase. How can a project leader take advantage of this for the project? First, the project leader can gather the team members together and gather lessons learned about what went on in the project. It is better to do this at the ends of phases than to wait until the end of the project. The knowledge will probably be lost then due to the elapsed time. Also, identifying the knowledge and getting agreement from the team builds teamwork and consensus.

Dealing with a Major Issue in a Collaborative Way

It is likely that the team will face a major crisis or issue in the project. For many of the most difficult situations, you have to rely on upper management for resolution and support. For other problems, have the team work together to address the crisis. The collaborative effort on the part of the team can help the project leader focus on potential actions and decisions.

Changing Project Direction

A project can change direction due to management action, external factors, or events in other related projects. The project leader should prepare the team for eventual changes in direction early in the project to avoid problems later. For example, the leader might propose several changes in direction based on detailed knowledge of the project. This allows the team to work with a reasonable hypothesis of change.

Action 5: Bring New Team Members into a Collaborative Environment

To bring a new team member into the project, include the following actions in a one-on-one session:

- Walk through the purpose, scope, issues, benefits, and other elements of the project concept.
- Indicate the history of the project and what changes occurred during the project and why.
- With this broad overview, review each team member and what their expertise is and what they do.
- Indicate how the new member will work with the existing team members.
- Review the issues with the new team member.

Next, introduce the team member to the project team. First state the new member's expertise and then explain how this person fits within the project. Indicate areas of joint work involving this new member and others. Have the new team member give some experience and lessons learned from previous projects.

Finally, set up and monitor the new team member's initial work. Each new team member should be assigned both individual and teamwork tasks. This will allow the new team member to experience a team approach.

EXAMPLES

An international bank wanted to implement credit card products across Southeast Asia and Australia. They decided to base the project in Singapore. A core team was appointed. All of the team members were based in Singapore. The banking organizations in each country were separately and individually accountable for their financial results. Bonuses and other rewards were based upon performance.

The project groups in all of the countries were told to participate by management. The management in each country assigned to the project junior staff members who had little bank knowledge. At the kickoff meeting in Singapore, attendees were told about the project and given the project plan. The plan for each country was basically the same. There was no recognition of the different cultures in the region. The team members were given no opportunity to provide input to the plan or to identify the issues of their individual markets.

What should have been a successful project turned into a nightmare. Staff members returned to their countries and went back to their own work. They generally felt isolated from the project. The project team tried to direct and dictate the project effort. The project failed after three months for the following reasons:

- Management of the bank operations in other countries fought the project and pushed for other projects that were more closely aligned to their profit goals.
- The core project team became demoralized because of the lack of cooperation.

The bank had no choice. It had to roll out credit cards to be competitive. Having learned the lesson of failure, the bank decided to try again. The approach employed was a collaborative one with a steering committee composed of business managers in each country. The project plan was created in a collaborative manner. The team members jointly developed the tasks as well as issues. Special attention was directed to cultural differences and the individual competitive situations in each country. This project was completed three months ahead of schedule.

GUIDELINES

- **As the complexity of the project grows, the benefits of collaboration grow.**

  The projects most unsuitable for collaboration are those that are small and short and performed in one location. For very large projects there must be a practical partitioning of the work among organizations and collaboration occurs at upper levels between groups.

- **Surround people who resist with people who participate and endorse collaborative scheduling.**

- **Work with the other managers to set priorities.**

  People are sometimes pulled off the team to do other work not related to the project. If you and other managers have jointly set goals, this problem will be less likely to occur.

STATUS CHECK

- Do you have an approach for doing work in a collaborative environment? Notice that the word *work* was used and not *projects.* It is helpful to have a pattern of joint work habits.

- How are projects managed that involve several divisions of your company? How are division-specific issues and problems addressed?

- How are projects managed that involve outside consultants and contractors? Do the consultants have any role in defining the tasks and work? Who evaluates the milestones?

ACTION ITEMS

1. Identify potential opportunities for collaborative work outside of projects. On projects, try to have about 30 percent of the

project tasks assigned to several people. Select a small project with several organizations involved and build a collaborative team. As the project goes on, attempt to gather lessons learned to improve your techniques for the next project.

CD-ROM ITEMS

10-01   Collaboration in Projects

CHAPTER 11

APPLYING INTERNET AND WEB TECHNOLOGY TO YOUR PROJECT

CONTENTS

11

APPLYING INTERNET AND WEB TECHNOLOGY TO YOUR PROJECT

INTRODUCTION

This chapter will focus on how networks and the Web can be employed beneficially in project management. These electronic aids are now an integral part of coordinating projects. This is especially the case with projects spanning multiple continents and time zones and with projects involving suppliers and customers. You gain a competitive advantage through intelligent use of the Internet and Web.

While you know what the Internet and Web are, intranets may not be as familiar. For the Internet and Web, your PC has Web browser software. This connects through the network to a local Web server. From there you access the general Internet through an Internet service provider (ISP). An intranet is your own internal Internet. An extranet links you with customers and suppliers using Internet technology. Both use the same software, but the data is contained within the company, without outside connections. What this means is that all of the functions associated with the Web and Internet can be supported internally on an intranet. So what advantages does an intranet offer? It is more secure than the Internet, since access to it is limited. Also, much of the new software targets this market, and it can be employed in support of projects. In addition, you can establish an intranet with suppliers, customers, and business partners.

Using the Web or an intranet with a Web browser in addition to the operating system offers substantial benefit for companies. First, they can house the software in a central server. Updates can be downloaded. This makes maintenance of the software and upgrades easier and less labor intensive. You don't have to upgrade each PC manually. Second, the PCs do not have to be upgraded with each new release of the software, since much of the software resides on the server. This extends the life of the PCs. A third benefit is standardization. If everyone uses a browser, training in software is reduced because many people have Internet and Web experience.

For example, suppose you wanted to employ project management software linked to a database of issues. In standard client-server computing, the project software as well as the application linking the database to the project management software would be resident on the PC. The database would be on the server. Whenever you clicked on a task, the software on the PC would issue a request to the server which would retrieve and download the issues. You would have to maintain the custom issues software on all of the PCs. If your setup and configuration files were tampered with, it is likely that someone would have to come out to your PC and do maintenance.

In the same example, if you employed an intranet, then the software and the database would be resident on the server. When you clicked on a task, the software on the server would be downloaded to the PC and the data retrieved. This requires more communications capacity in general but reduces support costs for staff time. It also encourages information sharing.

PURPOSE AND SCOPE

The purpose of this chapter is to look at how project management can benefit through the use of the Internet and the Web. Specific areas include the following:

- Collaborative work to perform project tasks
- Identification and resolution of issues related to the project
- General project communications

Guidelines are provided on how to get the most out of the Internet and Web without experiencing common problems and pitfalls.

The scope includes both project and project management documents and communications. Also covered are projects involving customers, suppliers, and business partners.

With respect to the areas of technology addressed, the chapter includes the following: the Internet, the intranet, the extranet, and the Web. Regardless of the technology, the functions needed to support project management are the same, whether internal or external. Thus, comments will fit both situations.

APPROACH

Project management and project work involve teams or groups. This applies to company projects and multiple company projects. Anything that requires communication within teams or groups could employ the Internet, intranet, and Web. Restrictions can be placed based on security and sensitivity.

The core applications of these electronic aids can be applied to general project information, project status information, project tracking, issue and action item information, and lessons learned. Other applications relate to the actual work in the project. Not only can you share technical and management information, but you can also support collaborative decision-making.

Example: Construction

The construction firm was able to spread the talents of the technical staff at headquarters to the projects in the field without the staff traveling to the remote location. Digital pictures were taken of project issues in construction and these were sent via the Internet. This was followed by Internet electronic mail and telephone communications. The manufacturing firm was able to install limited videoconferencing software at all PCs to facilitate interactive meetings.

Information can be shared across projects. The manufacturing firm was able to share information about government policies in Asia across the region. This was done previously to a limited extent but the Internet has made it much easier.

Web applications can be used in a project when a manager or a team is working with lessons learned, issue tracking, and action item tracking. You may also want to create the following items to place on the Web for your team:

- Static Web pages about the projects and project teams
- A contact list for specific issues or questions
- Dynamic Web pages for project updates

- Technical documents written within the project itself

Modern software supports exporting standard files into Web HTML files that can be accessed on the Web. These are feasible for the Web due to the number and capabilities of Web tools that are available. Some key features of the tools are the ability to generate, maintain, and enhance Web pages; the ability to attach graphics and images to Web pages; and the ability to connect a database or application to the Web page.

INTERNET AND WEB BENEFITS

The primary benefit is the worldwide accessibility of the Internet and Web. With so many people experienced in using the Internet and a browser, it is relatively easy to establish some project management applications. These applications are layered on top of existing applications and do not add to the learning curve of the staff. Since the marginal, variable cost of the Internet is small, communication is encouraged. When people are using the telephone, they may save up their issues and questions and then call another project worker once a week or so. This delays decision-making and issue resolution.

Some additional general benefits are as follows:

- Faster sharing of information
- Easier and wider availability of project information
- Visibility of information
- Easier training in the use of the tools
- Reduced cost in disseminating information
- Cost savings in communications and reduced travel costs

Here are some abilities of the Internet and the Web that you can use in projects:

- The ability to share documents between locations

  This has the potential to be a major application in the future. With improved multimedia systems and communications, you can have engineers or staff in different locations discussing and viewing specific problems and issues at a particular site. In the past this might have required travel to the location to address the issues.

- The ability to communicate regardless of the time of day

  If you use the telephone or fax machine, you often want to talk or send when people are in the office. This limits communications, especially when there is a 12-hour difference in time zones between two locations. Use of the Internet makes the communications more continuous and convenient.

- The ability to establish forms and databases

  These can be used to address issues, action items, and status.

- The ability to showcase information on products

- The ability to conduct electronic commerce

- The ability to perform a limited set of business transactions.

For the Internet, you are basically interested four features:

- Electronic mail

- Internet telephone

  The Internet telephone can be a useful way to increase communications due to low marginal costs. However, voice communications still have problems with response time and quality.

- Videoconferencing

  Videoconferencing puts an image on part of the screen. The screen image is refreshed at a specific rate of frames per second. The bigger the area the more communications capacity is required. The greater the number of frames per second, the more lifelike the video is. However, this also increases the communications workload.

- Files attached to the mail messages

  These files can be spreadsheets, project management software files, databases, or graphics files. This requires compatible hardware and software at both ends of the conversation.

ESTABLISHING THE BASICS OF INTERNET AND WEB USE

First, appoint coordinators. A project needs coordinators for both Internet and Web use. These coordinators should be someone other than the project manager. The general Web Master can support coordination for the Web. For the Internet, the coordinator should be a member of the project team.

Here are some duties of a coordinator:

- Support new users of the Internet in accessing the project information
- Monitor the use of the Internet for the project to ensure that it is not being employed for sensitive information
- Provide limited training and help support
- Provide for tracking of network problems and follow-up
- Monitor the Web to make sure that information is kept up-to-date

Once you have appointed coordinators, turn your attention to getting the rest of the team on board. For both the Internet and the Web, the key to effective use is to identify specific applications and uses of technology from the start. Explain up front how the technology is to be employed so that people understand what is expected of them in terms of use. Don't offer manual or paper substitutes, as people will continue to use paper and the telephone instead of electronics.

Next, take all of the applications covered earlier for the Internet and Web and apply them to your intranet, to provide centralized, secure applications.

After basic use and applications of the Internet and Web are established, the next step is to document continuing value from ongoing use. This is best done through organized updates of information in the application areas discussed. People must view the network as a key source of information and a critical way to provide and receive information on a timely basis.

DEVELOPING YOUR INTERNET AND WEB STRATEGY

What is a good approach for project management applications? The first step is to develop an overall strategy. How will your organization use intranets, the Internet, and the Web over the long term? Which of the applications mentioned will be most valuable? The answers to these questions depend on the company. Define a long-term list of possible uses, along with the benefits and your priorities. Determine how the applications group together. Here is a sample list of applications and some details:

Application	Type	Group	Comments
E-mail on status	Internet	1	Standard function
Transfer of project information	Internet	1	File attachment to e-mail
Document transfer	Internet	1	File transfer
Status of project	Web	2	Needs Web page
Contact list	Web	2	Static Web page
Action item mgmt	intranet, Web	3	Database application
Issues mgmt	intranet, Web	3	Database application
Document sharing	Web	3	Convert documents to the Web
Interactive project schedule update	All	3	Group access to project mgmt software is needed
Lessons learned	intranet, Web	3	Database application
Internet telephone	Internet	4	Needs compatible hardware/software
Collaboration on issues, topics	Internet, video-conferencing	4	Needs compatible hardware/software

How were the groupings selected? The first group is selected based on simplicity. No software development is required. The suggestion is to start with these applications; be conservative. Don't start more complicated development until you know that people will use the technology. They can then provide input for more involved applications. The second group is easy to do with Web development tools. The third group is the set of database applications for interactive project management approach. These applications provide real benefits and support each other. The fourth group requires specific software and hardware.

Group 3 items in the above table refer to project management applications. These are centered on databases that would reside on a server. Forms and data move from the server to the Web server down to your Web browser. Obviously, this does not happen automatically. Development related to the databases will be needed. Depending on complexity, programming of the links between the browser and the database may be required. If you want the applications to link, you will need additional development. The linkages support ease of use and access. Our experience is that if the links are *not* in place, many people will not carry out the links manually. To do so requires writing down the numbers or reference from one database, accessing another database, and then carrying out a search. This is a rather cumbersome process, eliminated by the creation of links.

Some suggestions of useful links are as follows:

- **Project tasks and action items**

  This allows a person to click on a task and see the action item information that pertains to this task. In reverse, you can move from action items into the tasks that pertain to this action item.

- **Project tasks and issues**

  This is similar to the above item.

- **Project tasks and lessons learned**

  Guidelines to assist in performing specific tasks could be accessible from the tasks. In reverse, tasks that pertain to a specific guideline can be obtained. This is particularly useful with templates which have the tasks, resources, dependencies, and resource assignment already in place. The access to lessons learned can assist in creating a schedule from the template.

- **Action items and issues**

  To address an issue requires action. If you have a number of action items to prioritize, then you want to know which issues pertain to which action items.

Group 1

With the first group of functions you can restrict your attention to templates, rules, and procedures, since only existing software and tools are needed.

- **Electronic mail on status**   The first step is to define a standard status template. If you receive status reports in a dozen formats, you will have a problem not only understanding them, but also trying to determine what action the information requires you to take. The template should contain the following:
    —The period of time covered
    —The scope of the tasks being reported on (reference here the tasks in the plan)
    —Progress and achievements on the tasks and major events
    —An update of the schedule and budget
    —Identification of issues and their status
    —Identification of action items and their status
    —Comments on upcoming tasks and expected results in the next period

    Also, set up a standard electronic mail letter that can always be accessed by everyone as a means of reporting on status.

- **Transfer of project information**   Rather than have the project information in the electronic mail text, attach files that have this information. The transfer note should indicate what the document contains, its format, and what the purpose of the transfer is for further action. Establish formats in advance that are acceptable for general use. You will need formats for project plans (project management software), documents (word processing), spreadsheets, and graphics files. Verify that the electronic mail system can accommodate the transfer in terms of preserving the document. Some computers that support electronic mail cannot support PC documents in terms of format controls. In such cases, people will have to export the file into a text file. The text file can then be attached. This causes more work for the sender and for the receiver, who must deal with the file conversion. The project team should adopt standard naming conventions for file names.

- **Document transfer** The same remarks apply as for project information. There is an additional case for transfer. That is, you can store some of the project documents on a file server. Members of the project team can then download the documents for viewing. In such cases, access to the documents should be protected so that read-only access is granted.

Group 2

Use Web pages for both intranet and Internet access. In the case of the Internet, the information should be treated as confidential and require password access. In addition, keep any sensitive information off of the pages. Assume that anyone can read them. Avoid photographs of the team and project stories. People outside of the project can get the wrong impression that too much time is being spent on the Web and not enough on the project.

Sufficient tools support generating Web pages and doing updates and maintenance. Little more effort is required than to do word processing.

Here are the two applications listed on the chart:

- **Status of the project** This Web page can be used to explain the purpose of the project, basic project information (scope, team members, etc.), and general status information on the project. You could also post the project plan in a summary GANTT chart.

- **Contact information** This is a list of project contacts. Information should include the following:

    —Name of contact

    —Position and firm, if external

    —Telephone number (office, home, cellular)

    —Address

    —Pager number

    —Fax number

    —Electronic mail address

    —Relationship of the person to the project

    —Areas of expertise

By area of expertise, include experts who have information and could provide assistance. You might include references to the Web master, project management software experts, other software experts, and technical and contract contacts, as appropriate. This list is useful since it can be found in a central, easily accessible location.

Group 3

Turn now to the applications in Group 3. These should be defined and prototyped to get reaction and ensure that you are providing sufficient value to encourage use. What is considered to be the incentive for use? Self-interest. If people see that it makes the job easier and reduces the overhead effort, they will tend to employ the software. Pilot the applications on the Internet. If everyone is located within one area, you can later establish an intranet. You generally cannot justify an intranet based on one project. Instead, decide whether an intranet is to be part of the methods and tools that will be employed to support all projects and line organization work.

How is the linkage established between databases and the tasks? In each of the databases (action item, issue, lessons learned, etc.), you can set aside a field for the task identifiers. Use this to refer to a task number in a specific field that does not change. If you refer to a task identifier that is the sequential number of the task, this could change as tasks are changed, new tasks are inserted, and tasks are deleted. In the project plan, use fields for action items, issues, and lessons learned. These will refer to the number you assigned to the items in the databases.

Here are some suggestions in developing these applications:

- Develop the database for each item. Design a simple screen for each database. Establish the linkage between the databases and the project plans. Then prototype the results through simulation. The objectives here are to ensure that the databases contain all relevant information and provide for ease in assessment. Load the current schedule and lists into the databases to add realism.

- After the prototyping, develop the links with the browser and Web. This will support the applications with manual linking. At this point you can test out the application on a sample project.

- Define rules and procedures for the applications and make them available, along with some training. Use the initial project to generate samples of workflow and use. Test these procedures with the initial project.
- After the applications have settled down, you can turn to developing the links among the applications and the project plans.

Group 4

The Internet provides for telephone and videoconferencing access and application. The first guideline is to evaluate and select the hardware and software carefully. Remember that all of the people must be capable of using the same tools. If you leave people out, you could isolate them from the project.

After selecting the hardware and software, install them on two local PCs as a test. Here you will shake down the technology in an environment that is local and can be controlled and monitored. Also conduct tests to determine benefits and to identify procedures on how to use the technology effectively .

The benefit of the telephone technology to project management is lower cost. The trade-off is a lower quality signal.

Videoconferencing can be a useful way to deal with issues and fact finding. It is not a good method for decision-making, due to the quality of the system (in terms of refresh rate in frames per second). If there is a heated argument or disagreement, videoconferencing can make people more frustrated. Use it only if you know that a major issue will not come up for discussion. For major decisions, nothing is as good as face-to-face contact.

Videoconferencing is more complex than standard telephone use. Not only must the sending and receiving units be compatible, but also sufficient communications capacity is required. In addition, rules must be established in terms of protocols of use. For example, how will figures and exhibits be employed and shown during a videoconference?

USING INTERNET AND INTRANET FOR MULTIPLE PROJECTS

Multiple projects that are related or that share resource cry out for the sharing of information. In all of the example companies, information on projects was shared through an intranet. Some examples of applications included the following:

- **Common database of issues** This allowed project team members and leaders to view each other's issues. This saved time later in the project because the project leaders could draw upon previous decisions.

- **Common database of lessons learned** These databases turned out to be invaluable because the projects within a firm were technically quite similar.

- **Project management software files** This allowed people to view each other's schedules without having to hassle people for the information.

- **Use of the intranet to allocate resources** The intranet was employed to allocate resources crossing multiple projects on a weekly basis.

USING THE INTRANET AND INTERNET WITH OTHER FIRMS

Prior to using an extranet or the Internet with any other firm, establish ground rules with team members on what materials will be sent via the Internet. Cover the format of messages and rules to be followed when people take vacations, are transferred, etc. This takes some effort up front, but it will save time and reduce the chance of misunderstanding later.

Information that we have found to be suitable for the Internet includes the following:

- Status of the project and plans
- Issues status, accountability, and comments
- Project team members and roles; how to contact people
- Lessons learned from the project
- Technical documents used in the project

Notice that we did not include personnel issues or budget matters. These are too sensitive to post on the Internet.

Some specific topics to cover with the managers of another organization are the following:

- Which transactions and documents are to be exchanged
- Frequency and timing of sharing of information
- Method of handling problems and disagreements
- Points of contact in both organizations

You can share additional information in a more structured way through Web pages. Include in your project team the Web masters of the organizations.

To coordinate a joint effort across multiple companies, establish a coordination committee or group that develops and implements the strategy. Benefits must be balanced so that both parties see it in their interest to pursue the implementation. For example, a major automobile manufacturer could force suppliers who did more than 75 percent of their business with the manufacturer to follow rules set up by the manufacturer. However, the company was totally unsuccessful with suppliers who did less than 25 percent of their business with the manufacturer. Thus, it is important to stress the joint nature of the project.

Example: Construction

Consider the construction firm. Prior to using the Internet, they faxed and telephoned daily to their key suppliers on projects overseas. With the time zone difference and work at the construction site, this was not satisfactory. They were often playing telephone tag. With the Internet they were able to use electronic mail to correspond on specific points about the project. Time zones were then a much less important issue. They then moved up to Web pages. Project plans, status, action items, and issues were exchanged with their suppliers.

INCREASING THE USE OF TECHNOLOGY

Once you start relying on technology, you can expect one of three outcomes. One is that usage takes off as people find creative ways to use the technology. The second is that usage diminishes and eventually stops due to a poor fit with the project or substantial resistance from team members. In the third case, use is limited to a few people and specific items. This is the outcome that needs your attention. How can use be expanded?

One suggestion on expanding use is to make available more information on the network for the project team. Make information easier to get from the network than from paper and faxes. For project team members who are resistant to technology, offer training and support. Try to find an expert on the team who can assist others.

In terms of suppliers or customers who are reluctant to use technology, you have fewer options. You can make it easier to use the network, but you lack the controls. You might consider what benefits you could point our or what favors you could give to those firms who adopt the technology. For example, on the case of retailing, suppliers will be paid faster with electronic commerce.

Security is regarded as a significant issue with the Internet. Even with encryption of data there are still concerns. Therefore, sensitive materials should be faxed or sent express mail or by parcel. Sensitive materials include strategies, personnel management items, and any political discussion. Even here, however, the Internet has a role. You can alert the receiver that a fax or package has been sent and provide a tracking number.

In your discussion of issues or problems, be careful about what you commit to writing. People can print it out and use it against you and the project later. Create a draft document and save it locally without sending it. Let the matter rest for a day. The following day, review it from the standpoint of someone who is hostile to the project. Then send it or hand deliver it.

Look closely at standards before you choose a technological product. Make sure that there is a standard that has been widely adopted for the technology. If such a standard does not exist, do not adopt the technology. You will pay too high a price when you later have to retrofit the technology to meet the standard. In terms of the Web browsers, use one release back from the latest version of the browser. The latest release often has more errors and bugs. The content on your Web pages should be compatible with most browsers for ease of access.

EVALUATING TECHNOLOGY USE—BEFORE AND AFTER

Prior to embarking on using the Internet or Web, evaluate the current communications process. Here are some specific items of information to collect:

- Dates, durations, and subjects of telephone conference calls and face-to-face meetings.
- Travel costs related to project work. Include lost time associated with travel, including getting set up for meetings and returning home.
- Elapsed time between the initial identification of an issue and its resolution.
- Comments about how the resolution and action item processes work in terms of difficulties due to misinformation, not having everyone involved present during discussions, and other similar problems.

After implementation of technology, evaluation includes these items as well as the following additional information:

- Contacts using videoconferencing, groupware, electronic mail, Internet telephone, etc.
- Trends in extent of use of the technology.

Signs of success in using technology are the following:

- Increase in electronic use over time. This indicates that people are becoming more comfortable with the new means of communication.
- Reduced elapsed time in resolving issues and in following up on action items. With a greater frequency of contact within the same period of time, there are fewer chances for misunderstandings.
- Lower fax volume, reduced travel, lower telephone costs, and reduced use of express mail and package services.
- More positive project attitude in terms of team spirit. People should refer to what other people on the project team are doing.

EXAMPLES

Modern Examples

The manufacturing firm used electronic mail with the Internet already. Thus, use in the project was a natural. Rather than employ groupware or Web pages, they decided to try videoconferencing. They were well suited to this because they had only three sites to link up. Thus, the cost was limited. Compatibility could be guaranteed. Moreover, two of the sites were connected with high speed communications already.

The first videoconference application was to discuss status and budgets. This was quite informative and people were able to see what was behind the numbers. Use of the videoconference expanded to issues and action items. Perhaps the most remarkable application was in providing expert help in project management from one location to the other locations. This started with project management and then expanded to the technical engineering side of the project.

The construction firm was a success story because people not only adopted the technology, but they also immediately applied it to project management. Travel, shipping, telephone, and fax traffic diminished. Issues were resolved faster. From an elapsed time of five days on average, the average time shrank to three days. This was due in part to technology and in part to the people involved and the methods they used.

Many cases have not been so successful because people became enamored with the technical side and did not effectively use technology for project management.

The consumer products firm enjoyed no benefits of remote communications since most of the people worked in the same general location. They employed the shared databases of issues, action items, and lessons learned. The results of the application were interesting to observe. It was relatively easy to set up and to get people to use the action items and issues databases. It was more difficult to use for lessons learned. People were not used to logging their experiences. Finally, the project leader set up an initial set of lessons learned as examples. The number of lessons learned then grew slowly.

Another problem in this firm was that people viewed the technology as overhead and did not see an immediate payoff. This was overcome when some of the results were shown to be useful.

GUIDELINES

- **Think of measurement and value constantly.**

  People may often question what you are doing with the technology and you want to be able to give an answer. In addition, you want to be assured that you are getting value for the investment. Follow the suggestions given in this chapter on estimating costs and performance before using the technology, as well as after.

- **Avoid linking to old, legacy systems.**

  In cases where you want to construct a intranet link with existing data in old systems, the interface effort can outweigh the benefits. This is generally the case when you wish to integrate old and new technology. That is why we suggest implementing the technologies we discussed with new hardware. Retrofitting and upgrading the old equipment can be too expensive.

- **Ensure security and protection of data.**

  You have two options. One is to have no security except standard log-on identification and passwords, as well as backup of data. Second, you can impose more stringent security and controls. This can be much more expensive and require resources to monitor and control the network. The project team works in the context of the organization and should follow the policies of the organization. If you have a choice, opt for less security and use more self-policing.

- **Run a pilot project.**

  Several valid reasons exist to run a pilot project. First, you want to verify that the forms, software, procedures, and policies are correct and complete. Second, you want to be assured that you are going to achieve the benefits of the investment. Third, you need to determine how much time and what amount of effort are required to learn the new technology.

- **Find other firms that have done similar things.**

  A number of benefits result from contacting firms that have experience with the technology you plan to use. First, these firms can advise you on what to avoid in installation and implementation. Second, they can give you hints and guidelines for use. Third, they can identify benefits that they have enjoyed. Having your higher level managers visit some of these firms can go a long way to convincing them of the benefits for projects.

- **Make sure of the compatibility of the tools.**

When selecting the software and hardware, be sure of compatibility with the network and existing hardware and software. You don't want an unpleasant surprise where an incompatibility surfaces that requires a substantial unplanned investment.

- **Build grassroots support for the technology.**

Technology is better employed with greater benefit when people *want* to use it, as opposed to being told by management that they *must* use it. Take advantage of the pilot project approach to do this.

- **Add as many functions to the intranet as possible.**

If you are going to employ an intranet approach, offer as much functionality related to the project and project management as possible. This will encourage use of a wider range of applications. People are more likely to get into the habit of using the software and network with more opportunities for use.

- **Have team members and others work with the technologies first in situations that are not important or stressful.**

A person working under pressure should not also be struggling to learn tools.

- **Before doing videoconferencing or even using electronic mail, get organized and be prepared.**

Try to use the technology efficiently. It is awkward to have to break off contact while you attempt to find pieces of paper and documents during a videoconference.

- **Set the agenda for the videoconference or group working session online ahead of time.**

  This is also true for conference calls. Stick to the agenda.

- **Use paper and notes to document the meetings as you go.**

  If you attempt to type the results while the meeting is going on, you will lose concentration and people will have to wait while you finish typing. The typing can be distracting and annoying.

- **Think compression.**

  When you are writing electronic mail or groupware messages, keep them short and focused. When you are transmitting files or data, compress the file in advance with a "zip" software package. During videoconferencing, have all exhibits ready. Do as little freeform drawing as possible, since it is difficult to read and follow.

- **Review the information to ensure that it is logical, appears complete, and is labeled correctly.**

  The work on the databases does not end with the establishment of the databases on issues, action items, etc. All of the software may have been established perfectly. The situation could fall apart because no one reviewed the information itself. Consider having each team member be responsible for reviewing new entries each week on a rotating basis. This will reinforce the need for accurate and usable information.

- **Review the content of the databases on a regular basis to see if the issues and action items mesh with the project plans.**

  Are team members keeping the information current?

- **Try to use the lessons learned in other projects or use them to change policies and procedures.**

  This will demonstrate their value to people and encourage more contributions.

- **Incorporate guidelines for using the technology effectively with the training.**

The purpose here is to have employees learn effective ways to work with the technology and tools, rather than being limited to the tools themselves.

STATUS CHECK

- Does your organization employ the Internet in projects today? If so, how has the usage grown since inception?
- What is the level of computer knowledge and expertise of your project managers? Of project team members?
- Does the organization employ groupware or shared project management software on a network?

ACTION ITEMS

1. For your organization develop a table that defines your intranet, Internet, and Web strategy. Here is a sample of the table. The column "Type" refers to whether the application is intranet, Internet, or Web based.

Application	Type	Priority	Group	Benefit

2. With the table in the previous exercise, discuss how the staff and managers would benefit from each group.

3. Construct a chart along the lines of the one in this chapter on benefits vs. effort of implementation. Discuss how this chart might change if you considered a large vs. a small project.

4. Assess the current technology available on your staff. Determine the level of staff expertise in technology.

5. For the project you defined earlier, determine how you would employ the following and what benefits you would expect:

- Electronic mail
- Internet mail
- World Wide Web
- Internet and internal web pages
- Groupware
- Shared databases on the intranet

PART III THREE

MANAGING
THE WORK

CHAPTER 12

MANAGING PROJECT RESOURCES

CONTENTS

12

MANAGING PROJECT RESOURCES

INTRODUCTION

This section of the book addresses the day-to-day management and direction of the work in the project. There are three parts to this puzzle:

1. Managing the **resources** and handling resource questions, issues, and opportunities.
2. Managing, evaluating, and coordinating the **work** and end products.
3. Measuring the **project**.

These three parts of management have different focuses. The first area, resource management, centers on individual and collective resources rather than on how to do the work and whether it is of acceptable quality. The second area, work management, focuses on the team. The third area, measurement, concerns an overall perspective of the project. Measurement provides the comfort and confidence levels desired in the project. In the big picture, resource and work management provide alternative perspectives of the project. Measurement is the umbrella for the two.

PURPOSE AND SCOPE

Resource management involves all project activities, including the following:

- Planning when resources are needed
- Putting resources together
- Setting up resources to do work
- Directing the performance of the resource

- Resolving resource questions
- Planning and staging when resources are no longer needed
- Releasing the resources
- Extracting lessons learned from the previous activities

Holding on to resources holds you back, since it costs money and effort to keep managing idle resources. Here is a lesson learned: it takes more effort to manage idle resources than busy resources.

The scope of resources includes employees, facilities, equipment, consultants and contractors, suppliers and customers, systems and technology, and external resources related to other projects. You do not have to manage all these on each project, but you have to understand how to plan and deal with each type individually and collectively.

The overall goal is to help you achieve the following:

- Get the most out of the resources
- Minimize the use of resources
- Minimize the time required by each resource

These objectives reflect some fundamental lessons learned. Many problems and issues in projects come from poor resource management. The fewer resources you use and the shorter time you need these resources the better. If managers are aware of what you are doing, they will be more likely to give you good resources. On the other hand, if you have idle resources, they may detect the waste and take less care in assigning qualified individuals to your project. Moreover, if you work toward minimization of resources, you put pressure on yourself to manage the resources more effectively.

The caveat to these objectives is to be aware of the political environment. It may be politically expedient and wise to waste some money and resources to achieve the overall project goals.

Example: General

A project manager resisted taking on a person who was politically well connected, but not a good performer. He finally overcame his better judgment and employed the person in the project for four months. It paid dividends in support, since he received more support from their organization.

Example: Construction

The construction company was already project-oriented, with tight cost control and an environment of strenuous controls. The project leader's problem was that, with success, his project attracted more people than were needed. The leader wisely took on several less qualified people, knowing that he would not get much from them. He did this as a favor to the line managers. They would realize that they owed him a favor. This strategy definitely contributed to the project by building political support.

The specific goals of this chapter are to help you with the following:

- Understand the timing of getting resources on board
- Reduce the time required for a resource to begin to work (cutting the learning curve for staff)
- Monitor the use of resources without micromanagement and interference
- Plan and implement the rules for releasing resources as soon as possible

Here are some questions that will be answered:

- How do you manage specific types of resources?
- How do you cope with commonly encountered resource problems?
- How do you plan the use of multiple related resources?
- Are you getting the most from your resources?

APPROACH

Step 1: Planning for Required Resources

Assume that you have developed the project plan and that it has been approved. Now you want to plan in detail for resources. Cut a deal with the various managers to get the resources. Concentrate on the resources required for the first few months of the work. Lining up resources too far in advance is a major mistake. The distant future is vague. You have shown no results in the project yet. You are not likely to obtain commitments for the best resources without a track record.

People Resources

Having identified what is required for the next few months, survey what skills are needed, what people are best suited for the work, and what people are available. You are now ready to negotiate.

Here are some guidelines for acquiring team members:

- Go to the managers who have the most available and the most easily obtained employees. Don't go after the most senior people, as this could cripple the line department.

  Who has the most resources? Consider managers who have the equipment, facilities, and tools required. Next, go to the support organizations that will play minor roles in the project. You are building momentum. You are gaining negotiating experience. You can use the managers here to help procure key people.

- Cut deals for the human resources that include a start date and an end date.

  The end date should include some allowance for slippage in the project. In negotiating, agree on ground rules regarding what you will do if you require the people for more time, if the project calls for their early release, and if the project is phased out. Convey to the managers your technique for measuring the effectiveness and use of the resources.

- Interview the prospective team members. Get them excited and motivated about the project. Explain the importance of their role. Be very specific on what is expected of them.

Material Resources

Consider last any resources for which you have to pay out-of-pocket money. This includes external purchases of equipment and consultants and contractors. It also includes work with suppliers and contractors.

Step 2: Getting the Resources on the Project and Up to Speed

A common mistake is that the project leader spends so much time getting the resources that he or she doesn't pay attention to the detailed activities necessary to get them on board. What does it take? For personnel, have an orientation for the project and the tasks that they will perform. For equipment, allow people to start using it right away. If a learning curve is required, plan for it. For facilities, you may have to prepare the space, install a telephone, and plan for parking. For tools, line up people who know the tools and have the necessary expertise. For outside purchases, start the procurement process. This is not trivial, since it can take months to go through the formal process. If you try to shortcut the process, red flags appear and you engender resentment. For needed consultants, do interviews and selection. If you neglect any of this, the project may suffer delays.

After a new resource is at hand, plan how to reduce the time before the resource becomes effective.

Step 3: Monitoring and Managing the Resources

Evaluate each major resource every day. Here are three ways to learn the information you need for an evaluation:

- Work side by side with the people doing the work.
- Sit in on meetings.
- Volunteer to do the work.

By actually doing the work of the project with the team members, you will learn the status of the work, gain the confidence of the workers, and gain information on any problems or opportunities that arise. If you follow up on what you discover, you will be a winning project manager.

How do people react when you become involved in the project work? At first, you may be met with reservations. People will wonder why you are there. When you show up repeatedly, however, they will be more open and friendly. You will be building an enthusiastic team.

When managing team members, get back to the line managers whose people you are using on a weekly basis. Give them an honest assessment of utilization and performance. When managing facilities and equipment, consider enhancing them so that when they are returned, they are in better shape than when they first entered the project.

Step 4: Planning for the Release, Replacement, and Termination

Now you are utilizing the resources. Start thinking about how you can continue to use the resource. In general, you want to release resources as soon as possible. Would the project benefit if you assigned additional tasks? If so, go to the line manager sufficiently before the termination date and explain the benefits of holding on to the people for a longer time than you first agreed on. Even if line managers require their employees, they may give you substitutes.

What is involved in release? Debrief personnel and get them ready for the transition. Prior to the transition, find out if they have any concerns. Take action on these concerns. Gather lessons learned and implement their ideas. Then contact line managers and work with them to determine how the employees' experiences can be used in their organization. You want the team members to capitalize on their experience so that their careers advance.

For the actual termination, gather the project team and announce that the person will be leaving and talk about what that person accomplished. Also, indicate what the person's new position will be. Invite the line manager. Pay a tribute to the team member for their good judgment.

This sounds logical, but what do you do when a project is spread out geographically? Try to go to the particular work site where a transition is taking place. You can gather lessons learned and then disseminate the information to the other sites.

Step 5: Implementing the Release, Replacement , and Termination

Implement the release as quickly as possible. If the time drags out, all of the good work you did in planning the transition is wasted. It is best to implement a release in the middle of the week so that the project members can establish a new environment without the resource before the week ends.

With the steps out of the way, let's turn to specific resource-related issues.

RESOURCE QUESTIONS

How Do I Address Morale Problems on the Team?

When you take over failing projects or suffer a reverse on a project, a common problem is bad morale. Why does this occur? The people often were not managed or motivated properly. They may have developed a negative attitude from an ineffective project leader.

This can be turned around. Get involved with individuals in the work. Build morale from the bottom up. Give people confidence in themselves by complementing them on the work. Get them additional support where appropriate and possible. Do not plan a group pep talk in the early stages. Get the team together only when you have begun to boost the morale.

If morale is good, don't assume that everything is fine. A team can get too confident. View morale as an ongoing issue. Keep building it up. Compare the project to your past projects and to projects in general. Never build morale by comparing your project with other active projects. If you denigrate another project, assume that the project leader will learn of it and this will reflect negatively on you.

How Do I Handle Problems with Team Members?

For political or other reasons you inherit or take on a team member who either has a poor track record or is not suited to the work. What do you do? You cannot just discharge the person. Go back to the project plan and examine what tasks are not assigned. Where could this employee do some good? Where could the employee do the least damage? Assign tasks and monitor the work, as you would do with a consultant. Ensure that this employee does not disrupt the work of other team members. Through your direct intervention you might be able to produce useful results.

If you try these suggestions and they don't work, try to persuade the person to leave the project on a voluntary basis. Don't criticize this person in front of others. If you are asked by team members what is going on, indicate that you are aware of the situation and are attempting to address it.

Here are some examples of problems with employees and how the project manager handled the situation.

- A team member was an alcoholic. The team member went through counseling, but the problem continued. The answer was isolation from the team until the person was removed from the project.

- An employee misrepresented his skills to the previous manager. It was obvious that the person was not qualified. A review of the entire project was instituted. It indicated that tasks had to be reassigned. This review led to the employee's removal.

- Another person appeared less than competent. However, it turned out that the person was salvageable through direction and guidance.

- An employee constantly criticized the project and the team. Morale was affected. The approach here was direct confrontation when the person was caught doing it. The person was reassigned.

After you have terminated someone from the project, tell each person on the team individually and answer any questions. Then hold a team meeting to go over what happened and the lessons learned.

Try to handle personnel problems yourself. Unless necessary, do not alarm management by reporting that you have a major crisis. If there is a crisis, focus on the issues and not on the personnel. This makes it easier to set aside emotions about the team member.

How Do I Cope with the Procurement Process?

Procurement is a cumbersome process that has been around since ancient Roman and Egyptian times. In Egypt there are records of the problems in buying resources. Roman officials constantly had to get approval from Rome to cope with problems beyond their scope. If Rome delayed decisions, the officials had to improvise with fewer resources. This was one of the factors in the fall of the Roman Empire.

Here are some guidelines for procurement:

- Go to purchasing and get a presentation on the procurement rules. Obtain the forms.
- Get a single contact person in purchasing to work with.
- Establish rapport with the person.
- Fill out requests for all of the resources you will need—even for resources in the distant future.
- Review all requests with the procurement contact.
- Establish a schedule for procurement.

Look at it from the perspective of procurement. Procurement staff often feel unappreciated. They get requests at the last minute. The requests are incomplete and incomprehensible. If you act in a cooperative and sympathetic mode, you will get more support from procurement. Offer to include the procurement person in meetings.

When procurement begins, support the process. Volunteer to interview potential vendors. Visit their facilities and talk to their managers. Keep the requirements specific and direct. Indicate the importance of the part or component to the project. Indicate how you will test and evaluate what is delivered. Update purchasing as you obtain and use the resources.

How Do I Manage Large Projects?

This is a curse that you should avoid. What if it is unavoidable? Divide the team into subteams. This will mean more work for you because you will have to function as the leader of a subteams as well as the project manager. Carefully design and craft how the integration will occur between work of the different subteams. Consider establishing separate schedules for each subteam.

How Do I Get People to Work Together?

People enter the project who did not get along before the project even started. Don't just force them to work together. Organize the tasks so that they have limited contact. To each person involved, acknowledge that the problem exists and that you will help minimize the effect on the project. When the people involved have to work together on a task, try to be there and participate. This will help head off problems. Work to keep them focused on the technical part of the work.

Example: Manufacturing

In a manufacturing facility, the design engineers fought with the production staff and supervisors. They blamed each other for every problem. The solution was to get them to focus on specific problems—one by one. A pattern of working through problems was established. The groups still did not get along, but the project succeeded.

Do I Manage and Direct Technology?

Many managers take technology for granted. They lack technical knowledge and treat equipment as if it were a mysterious black box. This is the wrong approach. While you don't need to know the details, all technology requires management. How is the technology being used? Is it being employed effectively? Do the people using the technology have the proper skills and training? Many times you will find that the use of technology becomes more effective if you actively assess how you are using it. Encourage the gathering and sharing of lessons learned. Do not be afraid to ask any questions that come to mind; as the project manager you are not expected to come to the project with a thorough knowledge of all the technical aspects involved.

How Do I Evaluate and Manage Consultants?

What are the potential benefits of consultants to a project?

- Consultants provide specific technical expertise that you will not require after the project.
- Consultants can fill a position when you are unable to find or hire the people with the right skills.

- Consultants can provide a different perspective and act as agents of change.
- Consultants can be motivated to do work quickly with the proper reward structure.

When you decide to work with consultants, follow the same guidelines given for procurement.

What is in it for consultants? They get paid and gain experience. Watch for less obvious goals that may threaten the project, such as the following:

- The consulting firm may try to take over the project.
- The consulting firm may use your project as a stepping stone to obtain other work.
- The consulting firm may try to keep adding people to your project.
- The consulting firm may substitute less qualified people on the project to increase profits, sending in the "A team" first and then gradually sending in the "B team" or "C team."

Some of the worst cases of abuse come from large client-server technology projects when the consultant firm puts in 20 or more people, which is too many at one time. Or, the consultant firm says it will supply certain people and then, when you need them, they are working on other projects.

How do you prevent these events? Recognize that these events can occur. Carefully think about what you require. Insist that you obtain certain individuals. If this is not possible, specify people with defined skills.

Here are some lessons learned about managing consultants.

- When evaluating consultants, don't just ask about skills and experience. Pose situations to them to which they have to respond. Test their creativity and common sense through this approach. Solicit questions from team members that you can use in the evaluation.
- Carefully analyze and work with the consultants when they first arrive. Micromanage them on the first tasks. After you have established a working relationship, you can back off.

- Play to the self-interest of the consultants. That is, indicate that you will provide a good reference and that you will try to help them with additional opportunities if they perform well. Don't go so far as to promise them anything that you may not be able to deliver, however.

- Ensure that the consultants transfer the knowledge and lessons learned to the regular staff. Insist on meetings in which they reveal what they learned. Work on the assumption that they will take what they learned and try to sell it to your toughest competitor. Exercise caution when using consultants, and debrief them often.

- Once they are established, make the consultants feel like part of the team. Don't differentiate between consultants and employees unless you have to for personnel reasons. You want to establish rapport and a relationship with the individual consultants. You want them to have a stake in the work.

- Have regular meetings with the consulting firm manager and review all work on a regular basis. Do this every week during a crisis. Move to twice a month if you are in a more routine mode of work.

- Don't wait until there is a crisis if a vendor is performing poorly. Contact the manager of the consulting firm with the details of the problem and the potential impact if it is not solved. Show that it is in the consultants' interest to solve the problem.

- Have very clear guidelines on when the work is to stop. Otherwise, the vendor may try to have the consultants on hand all of the time to run up the billings. Implement strong guidelines on vendor performance and enforce these.

- At the start of the work, indicate to all the employees what the role of the consultant is to be. The employees may feel resentful and jealous of consultants. They may feel that the consultants are getting the interesting work, which may be true, since consultant work is often creative. How do you handle this situation? Tell the employees that their job is to learn from the consultants so that they can do the work in the future themselves without the consultant.

- Do not be defensive if the consulting firm tries to run the project. They may start by volunteering to help in planning. To deal with this, get material from them and incorporate it in the plan yourself.

- Go directly to management and find out what is going on if the consultants go behind your back to management. Before they come on board the project, indicate to management that this may happen and explain how you are going to manage the consultants. Ask managers to come to you right away if they have any concerns.

- If a consultant is called in by a manager to audit or review your project, carefully pin down the scope of the review and the objectives. Verify this with management. Provide information to management within the objectives and scope only. Don't whine about what impact this has on the schedule. Don't appear defensive. Inform the project team that management thinks that the project is important or has concerns and wants the review. If you show limited cooperation, the potential damage will be limited. Use your project knowledge and your understanding of the political structure to your advantage when you talk with management.

How Do I Coordinate Multiple Consultants?

Take the previous problems with consultants and compound them with multiple consultants. This occurs in large projects with tight deadlines. The manufacturing project manager faced this situation. She had to coordinate consultants in each country along with overall consultants across the project.

Here are some of her lessons learned.

- Play one consultant against the others with regard to issues. Each consultant will act from his or her self-interest and perspective. This will yield better solutions to problems.

- When addressing an issue involving multiple consultants, get all of them in a room with the staff, if possible, and sort out what can be done.

- Establish in the meetings exact tasks that each vendor will perform. Establish a coordination process.

If one vendor criticizes another vendor, get them together and state the concern. Don't show favorites. If you know that you have a poorly performing consultant, use the other consultants to provide data to justify terminating the first consultant.

How Do I Address Non-Performance and Schedule Slippage?

When you detect signs of this in advance, pin down the cause of the slippage. Here are some potential causes of nonperformance:

- Some tasks were not in the plan.
- Quality problems indicate the need for rework.
- The project is not receiving the resources on schedule.
- The team is not performing the work on time.
- Time is wasted on the project.
- The tasks are vague or ambiguous.

Some of these situations can be prevented through front-end planning. If the problem arises, go to the team and see what can be done to address the problem with current resources. This will reveal the extent of the problem and what is going to be required in terms of management patience and support. If you have to choose between asking for more money or more time, choose time. Adding money means adding resources that could further delay the project.

As you are doing this, alert management through informal channels that an issue has come up that could affect the schedule. Let them know that you are working on the problem and will get back to them. Gradually unfold what is going on to management. Give management options as to what to do with your recommendations. You want them to buy into the decision and become supportive.

How Do I Coordinate Work With Suppliers and Contractors?

With downsizing and new industrial relationships, more projects involve customers and suppliers. For example, automobile companies work with Electronic Data Interchange (EDI). This has led to some spectacular successes and some dark failures. Success occurs when the external organization sees the project as being in its own interest. In cases where self-interest is not evident, the supplier or the customer is less willing to participate in the project. This is natural.

So, before you start, identify this self-interest for the supplier or contractor and capitalize on it from the beginning. Continue to stress these benefits as the project progresses.

What do you do when the supplier or contractor lacks the necessary skills or expertise required by the project? To prevent this from occurring, develop a survey form and employ this to select the right supplier or contractor. If this is not possible, first carefully define the project's necessary tasks. With the detailed project plan in hand, go to the firm and review what end products are needed. Then move to defining how the project will interface with their people. This will lead into a natural discussion of the roles of the people from the firm in the project. Before the people are named, review the skills, knowledge, and degree of involvement that will be required. Define a review process that will be employed to coordinate between the two organizations.

You often depend on the work and schedule of other projects. They may depend on you. A suggestion is to coordinate this like you would the supplier and customer situation above.

Here is a list of some of the activities that will help you coordinate work with suppliers and contractors:

- Identify the dependencies between the projects.
- Review the exact content and standards of the deliverable items between projects.
- Define how quality control will be addressed.
- Define the coordination process between the projects at and below the level of the project leader.
- Determine how project schedule and priority changes will be communicated and addressed.
- Define the escalation process for moving disagreements and disputes higher in the organization.

EXAMPLES

The Railroad Example

On the railroads, supervision of personnel was delegated to the managers in the field. Efforts were made to achieve some consistency. This was difficult, given the size of the project, the distances, the limited technology, and the small number of managers. The best solution that evolved appeared to be the rotation of people so that they could move periodically to new roles with new managers. This had the positive impact of tempering the bad managers.

Modern Examples

The construction firm had little or no trouble with the non-personnel parts of the project due to their line of work. The problems came in interpersonal skills and coordination. Long-standing conflicts between project managers hampered the work. Some project managers acted like prima donnas. This was one of the project leader's major challenges. He had to intervene regularly between different project managers. The project managers would not openly fight, but they would become quiet and nonresponsive. The project leader first thought that silence meant progress. He later realized that it could mean resentment and lack of work. His technique was to gather information from the various leaders, work out a solution, then meet with each leader individually and present the solution.

The problem in resource management at the consumer products firm was in getting participation from both the schedulers and the product managers. The manager settled on a technique in which each problem or situation was defined and then interpreted from the standpoints of both the project manager and the schedulers. In some meetings he acted as an interpreter.

GUIDELINES

In Chapter 5, some resource issues were addressed. Additional guidelines will be added in this chapter from the perspective of overall management of the resources.

- **Understand as much as you can about your consultants.**

  Motivation has already been mentioned. Try to understand their culture and how they reward their employees.

- **Give rewards for work, but avoid financial awards.**

  Financial awards are often counterproductive. People will start expecting them and take the money for granted. Money is probably not the right incentive. As an alternative, give time off, which has the added benefit of preventing burn out.

- **Don't allow an issue to fester.**

  An issue can arouse passion. This is good if it is directed toward the resolution of the issue. If the emotions aren't channeled, the team may divide into warring camps. This is difficult for a manager to overcome, since the manager would need to find middle ground. If that happens, the best plan is to work on other aspects of the issue to get it resolved.

- **Make sure that people maintain awareness of their home organization.**

  In a project of substantial duration, some people on the project can become hangers-on and stay too long on the team. This happens when people have been on the team so long that they lose touch with the line organization that serves as their home. Encourage people to visit their friends and their line manager. If you know that the project is going to last a long time, start this policy at the inception of the project.

- **Use the meetings on issues to allow people to clash and argue about methods, tools, and approaches.**

  Clashes in a project team are expected. Carefully direct these meetings to be fair and do not allow the senior people to dominate the meeting. Younger team members can contribute new ideas and hard work and should not be overlooked.

- **Avoid adding new people to a project team.**

  If you add anyone to the team, they have to become acclimated to the team and the work. They have to be socialized into the project. This slows down the progress of the project.

- **Monitor part-time project people.**

  Managing part-time people on the project is a challenge since their time is spread across many projects. Furthermore, you do not have full control. How do you get them to continue to work on your project? First, follow up on what they are doing. Set the duration of their tasks and level of detail of work so that you can monitor it more easily.

- **Always have junior, less experienced people on a project.**

  These team members are often energetic, intelligent, and eager. They may lack experience, but their more recent training and knowledge may be valuable assets. The junior people are more open to change and new ideas.

- **Periodically review the roles of each team member.**

  You have defined the roles of the team members (see Chapter 5). However, the project goes on for months and people forget or lose sight of their roles. Not knowing the roles, the team members invent them. To avoid such problems, regularly review the roles of each team member to the group. Otherwise, you risk losing control of the work.

- **Deal with conflicts in an open manner.**

  External organizational conflicts will be reflected in the team. Get the people involved in the conflict in a room and acknowledge the conflict. Indicate that the project is a different entity and that they will need to work together effectively on it. Reinforce this whenever problems arise.

- **Pay particular attention to the timing of a resource change.**

  Timing of changes is important—sometimes more important than the change itself. You may want the change to occur in a critical period to shake up the project. It is not necessarily true that the best time to make changes is when everything is calm.

- **Be flexible on controlling the team members in terms of their time.**

  Managing the team by the clock will yield presence but probably not results. If people are doing project work at home and you have evidence of this, tread lightly.

- **Cultivate your power of memory.**

  When managing resources, take notes and reinforce your memory. You want to be able to relate events to each other over time.

- **Carefully consider team member substitution.**

  Substituting project staff can cause unforeseen ripples. The person who is being substituted may have developed personal ties and may have performed work of which you are unaware. Investigate to see what the impact of substitution might be before making a decision.

- **Remove people from a large project to speed up the project.**

  For large projects that are in trouble, consider removing resources from the team. This can be carried in conjunction with narrowing the scope of the project and restructuring the work.

- **Be careful when you shift resources between tasks.**

  Moving resources between tasks can slow down both tasks. People have to transfer what they know to their successors. They also have to pick up information from their predecessors. Shift resources only as part of a major project change to minimize the overall disruption.

- **Keep the overlap of people entering and leaving a project to a minimum.**

  Overlap is good in that you can transfer knowledge. However, it is best if this is short. Otherwise, it is awkward for the person entering the task. Bad habits and misinformation can also be transferred.

- **Nurture relationships with team members for later projects.**

  Try to get acquainted with the people who are performers on the project. You may be seated next to the same person on the next project.

- **Allow people to express emotion.**

  People on a project team can become emotional on a wide range of issues. This is not a problem in itself. Allow people to vent some of their frustration or happiness relating to the project.

STATUS CHECK

- How are personnel issues addressed in the project? Are they largely ignored until there is a crisis?

- What amount of time is consumed by idle resources? What is the impact of idle resources on the productivity of the team and project?
- How many missing tasks exist in planning, in setting up, in getting ready, and in shutting down resources?

ACTION ITEMS

1. Conduct an assessment of how non-personnel resources are being managed. Determine the impact of resource problems on the project.

2. For personnel resources, identify the interpersonal problems that exist within the team. Has anything been done about these? What steps do you think can be taken now?

3. Return to the project plan and add tasks that include procurement, setting up resources, and removing resources. What is the effect on the overall schedule? If the result has a major impact, you are missing a significant amount of work and your schedule is probably not realistic.

CHAPTER 13

MANAGING THE PROJECT WORK

CONTENTS

13

MANAGING THE PROJECT WORK

INTRODUCTION

So far you have constructed the plan, started the project, and managed the resources. All of these things are important, but the bottom line is the work in the project—how to set up for work, perform the work, improve the work, and evaluate the results. As a manager, you want to be able to evaluate the work and improve the work where possible. When evaluating, many managers immediately consider the most routine, common work performed in the project. Experience shows, however, that problems are more likely to occur in areas of exceptions, rework, workarounds, and nonrecurring work. Mistakes are more likely to be made here. Estimates of resources required and the schedule are likely to run into trouble here as well.

As an example of nonroutine work, consider workarounds. A workaround occurs when you have a problem with some aspect of the work, such as tools, methods, systems, facilities, or even certain people. In order to get the work performed with the available resources, you have to invent a way to get around the problem—a workaround. If you manage routine work and ignore workarounds, you are ignoring a substantial percentage of the work. Instead, spend time managing the workaround. Once you turn your attention to a particular workaround, the best tactic might be to try to eliminate a workaround by backing up and addressing the specific problems that created the need for it in the first place.

PURPOSE AND SCOPE

The purpose of this chapter is to determine what is going on in the actual work and to consider how to improve the work. Improving the work will benefit the schedule by reducing uncertainty.

The scope of what is going to be considered includes all aspects of the actual work, from set-up through performance to shutdown or completion.

APPROACH

First, we will examine which tasks to consider for analysis. We will cover how to review these tasks. Ways to improve the work are offered. Specific situations are then analyzed.

Step 1: Choosing Which Tasks to Analyze

You don't have time to check all tasks. When choosing which to evaluate, the obvious place to start is with the ones that are in trouble. To know which ones to check after that, assess risk. Risk is examined by considering the issues underlying the risk and estimated by looking at the likelihood of slippage and the problems that will result if slippage occurs. Tasks that have moderate likelihood of slippage and exposure are often more dangerous to your project than something that has high likelihood of slippage but low exposure, or vice versa.

Review your list of active and soon-to-be active tasks and first determine the importance of each task to the overall schedule and project. Take into account more than the mathematical critical path. Also consider the potential side effects on other tasks and projects—the ripple effect. This process will give you a list of the tasks with some amount of exposure.

Next, think about the people doing the work. If necessary, do a brief on-site review of the work and determine which have some likelihood of slippage. Look for tasks that have exceptions, workarounds, and rework. The times for these things are less easy to predict. After evaluation, you should be able to determine which tasks have the greatest risk of slippage.

When looking at which tasks to analyze, separate out the exceptions. Look for instances where the routine work (which is likely to be the most efficient) breaks down.

Another area to investigate is nonroutine or nonrecurring work. These are tasks that involve more thinking. People will tell you that these cannot be planned or worked through quickly. However, you will find that many of the subtasks and detailed work are similar to standard work.

An important area is to consider where and when work is turned over between people. This occurs between shifts, or when people are going to move on to another task. Many things can go wrong in a hand-off. Often, the time allowed is too short. Also, people may write cryptic notes and just take off from work. This then necessitates telephone calls to get status. When one worker requires information from another, both can waste time with the telephone tag syndrome.

Step 2: Evaluating Work Results and Milestones

Once you have determined which tasks to evaluate, rank these tasks according to levels of review, as follows:

- Level 0—No review
- Level 1—Evaluate for existence of work
- Level 2—Evaluate for presence of the work
- Level 3—Assess the content of the work

Levels are helpful in evaluating work because it is difficult to evaluate all the work at Level 3. For some tasks you will take people at their work that they completed the work. Differentiate between existence and presence. Existence just considers that the work is there. Presence checks the form of the work as well as existence, but it does not address content. Level 3 reviews and analyzes the internals of the work. Decide on the level for a milestone based on risk and importance to the project. Many milestones must be evaluated at Level 1 due to the number of milestones and the limited time and resources available.

Once you determine the milestones for detailed review, you can identify who will participate in the review. These people should be familiar with the project; they can be other members of the project team. However, if the same people are always used, the quality of the review will degrade over time.

What do you hope to accomplish in the review? First, determine if the work product is of sufficient quality so that you can continue. Second, transfer and share knowledge between the reviewers and the people who did the work. This helps build a team and can assist in providing backup later if someone on the team who did the work is no longer available.

Should reviews be planned or unannounced? Planned reviews have the advantage that materials will be available. However, in a dynamic setting, the project leader and the team have to be ready for a review at any time.

How should a review be structured? It depends on the nature of the milestone. The project team members who performed the work may present the milestone without intermediaries. A sequence of steps might be as follows:

Step 1: Hand out materials for review in advance of a meeting.

Step 2: Set the agenda for the meeting by posing specific questions and issues about the milestone.

Step 3: Conduct the review of the milestone.

Step 4: At the end of the meeting, determine what is to be done and what actions are to be taken.

The project leader should orchestrate all of these steps and also act as a scribe during the meeting. The review should concentrate on the milestone. Bring up the details of the work only if there is a major problem.

Possible results of the review are as follows:

- The milestone is turned back as totally unacceptable; the project stops.
- The project will continue but the team will rework the milestone.
- The milestone is acceptable, but additional work will be done later.
- The milestone is acceptable but additional tasks will need to be performed.
- The milestone is acceptable; no additional work is required.

Now expand your review to how the team members are supervised. Here are some areas to consider:

- How team members are supervised on an hourly or daily basis
- How team members are assigned work
- Preparation time is allowed for the work
- Problems that team members have with coworkers
- How reported problems are handled by supervisors
- How supervisors accept and follow up on suggestions for improvement
- How work is checked and evaluated
- How supervisors respond to problems related to equipment, parts, and supplies

Working with supervisors to change priorities is difficult and may require intervention from a line manager.

Now review your spending of money. Spend money in chunks so that you can trade off among different spending candidates. If you spend money on one item at a time, your money and people's patience will run out. Give higher priority to action items that involve one-time expense rather than recurring expense.

Step 3: Improving the Work

Where should you start to look for improvement? Consider all activities that don't require money, organization approval, or involvement. Ask the people who are doing the work what they would like to have happen and change. Test out simple, inexpensive, and non-political changes.

Here are some examples of such changes from past projects:

- Relocate where the work is performed
- Relocate support equipment
- Rearrange the work layout
- Clean up the area
- Service the equipment
- Upgrade the procedures with the new versions

- Have people share experiences
- Fix and repair equipment
- Obtain updated procedures
- Train staff in policies and procedures
- Make small procedural changes
- Obtain small, inexpensive parts and components that make life easier

Your best sources of information are direct observation and the people who are doing the work. Examine how exceptions and rework are handled, since these can consume much more time. Start with the question, "If you could do things differently, what would you like to do?" This should solicit responses that match some of the items on the previous list. Next, start to implement changes. Morale will climb.

Here are some action item candidates for improvement:

- Upgrade equipment or software
- Upgrade facilities
- Upgrade skills
- Add on to the support
- Add more people
- Replace equipment
- Replace facilities
- Consider outsourcing or insourcing
- Replace the people
- Change the process around the work
- Increase the extent of testing and quality of testing

For each idea, determine costs associated with acquisition, installation, training, set up, adaptation, modification, usage, and management. Once you have groups of changes, you can determine the overall impact. You can change the schedule to accommodate groups of changes.

HOW TO RESOLVE WORK QUALITY ISSUES

Situation 1: Work Quality Is Poor

This task may have been in trouble from the beginning, but other explanations for poor quality are possible. First, the measurement of quality may have changed—it would have passed three months ago. Second, the standard of measurement may have changed. Third, work quality itself may be acceptable, but the appearance or impression is not acceptable.

Look for causes and effects. If problems are due to one of the first two situations, management changes in direction are a strong possibility as a cause. Personnel turnover, new equipment that is not calibrated, new procedures that have not been shaken down, the work being speeded up, and morale problems are other possibilities. Consider combinations of events in a limited period of time that may affect the project.

After you have determined possible causes for the problems, answer the following questions:

- What event suddenly caused the work to be unacceptable?
- What problems have been gradually getting worse?
- What is the effect of poor quality on the project? Can the project continue?

Interview the staff to find the source of the problems and the interaction between the causes. After finding a likely source, run a pilot test to determine what effect improvements could have. Look at related areas. Has the problem spread?

After your analysis, you can make an effort to improve the process overall.

Situation 2: It Wasn't What Was Planned

Surprises are part of the life of a project. Perhaps the purpose of the project changed in a meeting, causing half of all of the project work to date to be superfluous. Or, the scope keeps expanding. It seems to grow like a crop of wheat. Or, you thought that the task a team member was working on was going to take two weeks. Two days into the task, the new estimate is four weeks and climbing.

Here are some steps you can take when you encounter surprises:

Step 1: Gain perspective

A first response to situations such as these is to sit back and get an overall perspective of what is happening. When you modify the purpose, scope, or work, you will add tasks. Make an effort to do this immediately. This will test your understanding of the changed conditions as well as provide you with information for the new schedule.

Step 2: Determine how it happened

This is not intended to place blame. You want to examine in retrospect why the situation occurred so that you can improve your estimation and management in the future. Could it have been predicted if you had watched the work more closely? This step will sharpen your monitoring skills in the future.

Step 3: Seize the opportunity

Use this situation as an opportunity to see if the work and the work environment can be improved. Can you eliminate or combine any tasks? Can dependencies be changed? Can the quality or quantity of assigned resources be reduced? Implement changes while the situation is fresh.

COMMON TYPES OF PROJECTS

Here are four types of projects. Information is given on how each is managed.

Construction Projects

Construction projects long ago ceased to be simple, self-contained projects. Multiple contractors and subcontractors are involved. Marketing of the finished product is important. The permit, environmental impact, and planning processes take much longer and cost more than they did years ago, generating the need for even more capital.

Managing a construction project is a multilevel effort. At the site there is day-to-day and hourly direction of the work. Moving up a level, planning and staging of materials for the next set of tasks is necessary. Higher up is the management of the contractors and subcontractors. The fourth level is the overall project level. Some projects have even more levels.

What is the critical focus for managing the work? It is partially the individual levels. Mainly, however, it lies in the coordination of the levels. For example, you are dealing with a contractor problem in the third level and this affects the lower levels as well. Or, problems at the site must be worked up to higher levels quickly before the schedule is affected.

Product Development

The consumer products firm has the classic product development situation. Because the products are new, each manager considers the work to be nonrecurring. Each product then tends to have a unique product management plan based on the style of the manager. Managers feel autonomous since each one is being held accountable. Managing the work across the product development projects is a challenge, because the projects are developed individually.

How do you treat nonrecurring work? The first step is to break it into components. Once divided into units, some tasks will reassemble and become recurring. This narrows the field of nonrecurring work. For a nonrecurring task such as the design of the product, aspects of the design still follow a pattern. At the core of the firm's project is the standardization. Nonrecurring work can be placed into a recurring structure since it is only the content and detail that is nonrecurring. Durations, dependencies, and other tasks remain as usual.

Software Development

Software development is a fuzzy area. Only when you can test the results of programming with tangible computer code do you know what you have. You do have requirements, design, and specifications, but these exist only on paper. That is why most modern software development now employs prototyping, so that people can see and touch mock-ups of a system.

How do you manage the work of the developers on a project? One suggestion is to keep the project team small. Divide the project into small tasks so that milestones are defined frequently. Conduct work reviews and meetings with programmers and analysts to see what they are doing and assess the quality of their work.

Software development management gets more complex in integration. Determine how to integrate the software programs and how to repair errors that are found, all at the same time. It is a moving target—fixing, testing, and integrating. Most large software development firms have turned to the method of successive builds of the system. In this method, the software is integrated on a regular basis and then tested weekly. When the programmers are working on errors from release 23, the testers are working on release 25 and the integrators are putting together release 26. The method of successive builds of software is like a series of dress rehearsals for a play.

Technology Implementation

Deploying technology is no longer a simple task. Multiple vendors are involved. There is integration of subsystems. Testing and analysis of results must be undertaken.

Here are some things that can go wrong:

- The vendor products do not interface
- Vendor staffs do not share information with each other
- You are always waiting for the latest fix to the latest problem that you encountered
- You may be the only one who has attempted to integrate these very specific products into a system

How do you manage this work? Vendors should have their own plans. Then you can integrate the overall plans. Define dependencies between tasks so that when work in a specific task is being discussed, you will be aware of the other work and tasks that are going on concurrently.

EXAMPLES

The Railroad Example

The most common management technique was on-site management of the work. The target of the schedule was always uppermost in people's minds. You could get priority with resources if you could show that schedule improvements were possible. Quality standards of work were limited. It was acceptable to have substandard track bed or bridges because people would argue that this would be fixed after the line was completed. The argument was that there would not be much traffic on the road until after the problems were fixed. This was an attitude of "do it now, fix it later."

Modern Examples

For the manufacturing firm, managing the work at the remote countries was delegated to a project manager in each location. To increase consistency, this was later augmented by having an oversight consultant go out into the field and review the work as it was being performed.

For the construction firm, each construction project was carried out in a similar way. Managing the work translated into identifying situations in which risk was increased due to local conditions. This could be due to weather, local labor conditions, the work site, the quality of building materials, and other local factors. Management of the risk and work then meant identifying tasks and following up with direct, on-site attention.

Product managers at the consumer products firm directed the work in a hands-on mode. This was the best approach for internal work. However, many of the products involved contractors and subcontractors. These had to be managed by having the product managers visit the sites. The most successful product managers were those who undertook frequent site visits, some of which were unannounced.

GUIDELINES

- **Be selective in deciding which areas to enter as a project manager.**

In some cases, if you interfere in the work, you can create more problems and issues than you address.

- **Formulate acceptance and rejection criteria for milestones.**

  Work is driven by milestones and end products. Without a detailed understanding of what is expected, it is difficult for team members to structure, perform, and measure the work. The project then moves out of control.

- **Know whether to close off alternatives or keep them alive.**

  In the day-to-day work of a project, hesitation can cause work on the tasks to freeze. People say that they cannot do the work without the information or a decision. Therefore, make the effort to close off alternatives to maintain momentum.

- **Be a hands-on project manager.**

  Attempting to manage a project remotely can lead to isolated results. Some managing can be performed remotely by probing into how work is being done, but the requirement to be hands-on and see and touch the work is still foremost. If you decide to manage tasks through delegation (which means remotely), you are only going to have pockets of success, achieved when you are working with experienced people who have clearly defined tasks.

- **Concentrate on the result, not on how work was done.**

  In reviewing a milestone, it is often tempting to ask why someone did something the way it was done. This leads to probing into how the work was performed. The problem here is that this detracts from the milestone itself.

- **Periodically examine some of the irrelevant but interesting project tasks.**

  People often ignore these tasks, thinking them too mundane. However, if you could improve the work and schedule of these tasks, you could devote more time and resources to the more risky tasks.

- **When measuring success, count on your own opinion.**

  Work success can be measured in terms of schedule, cost, and quality. But these are sometimes difficult to measure. Some managers rely on the assessment of team members. This is dangerous since it can be misleading. Also, you are depending on others for their assessment. Measure work by what *you* did or by what effect *your* work had on the organization. Do not depend on the organization implementing and using the project results immediately. Actual use may be a separate project.

- **Focus on achievement rather than work.**

  Work is not the same as achievement. If people tell you they worked hard, what comes into your mind? Do you assume that because they worked hard, they accomplished a lot? If so, you might be mistaken. Ask what they accomplished.

  To achieve means that people organized the tasks, performed the tasks, and got the desired results. Doing work means only that they performed the tasks.

- **In most cases, work problems are best turned around without new, massive amounts of resources.**

  Adding resources can demoralize the current resources. Compounding this problem is the additional time required to bring the new resources up to speed.

- **Get work from each person on a regular basis.**

  Review samples of what people are doing. This shows that you, as a project manager, care. It shows that you are interested in the project all the time, not only when there is a crisis or problem.

- **Change purpose and scope if the schedule is stretched out.**

  If a schedule slips several times and no one takes any action or attempts to adjust resources, people will stop taking the schedule for the work seriously. Work within routine tasks may slip. The project slowly moves toward failure.

- **Monitor tasks people are working on to make sure that the more difficult tasks are not being avoided.**

Some people work on easy tasks to build volume, while the risk and importance lie in more difficult work. It is human nature to perform many simple tasks first, to show that something is being accomplished. This psychologically can provide momentum to tackle a difficult task. However, it can be misused to avoid the difficult work.

- **Consider the degree of uncertainty in estimates of work.**

  Slack in a task is the time that you have before the task begins to affect the overall project schedule. Slack is a matter of interpretation. If you start with slack of five days on a five-day task and after one day the slack slips down to three days, you are encountering uncertainty about the work. The amount of slack in a task depends on the degree of uncertainty. When you review the work in a task, try to get an understanding of how a person determines whether he or she is definite about estimates of work.

- **Look for fuzzy tasks.**

  A fuzzy task can usually be identified by its wording and weak assignment of responsibility and resources. Hidden behind a fuzzy task is typically a set of undefined, yet-to-be-determined, detailed tasks. Make the poorly defined tasks more specific by adding detail.

- **Hold one person accountable, even if several are involved in a task.**

  Separate who is doing the work from who is accountable for the work. Have single-point accountability for each task.

STATUS CHECK

- How do you measure your own work in a project? What quality standards do you attempt to meet?

- To what extent is an effort in place to define tasks with risk and tasks to which project management attention is directed? Or, alternatively, is each project manager left to his or her own devices to figure out how to review the work?

- How is the management of nonrecurring, exception, workaround, and rework effort managed? Is it handled differently from routine tasks?

ACTION ITEMS

1. Practice analyzing work. Take a specific task that you or a friend perform. Divide the task into component parts, as discussed earlier. Define several alternatives for improving the work. How would you implement the changes? How would you measure the results?

2. If you are in a large organization, how are milestones reviewed? Is the process different among projects? Do you use standardized checklists? How often are end products turned down as not being acceptable?

CHAPTER 14

ASSESSING PROJECT EFFECTIVENESS

CONTENTS

14

ASSESSING PROJECT
EFFECTIVENESS

INTRODUCTION

This chapter is concerned with measuring the work in the project.
Measurement is taken in its broadest sense. That is, you are interested
not only in understanding what is going on, but also why and how it is
happening, so that you can make informed decisions. Measurement is
an area that receives little attention in project management. The
traditional approach concentrates on milestones. Measurement of the
effort is left to budget vs. actual and earned value analysis. We will
examine the modern day versions of these.

Why is measurement important? It makes marketing and selling
possible. It requires you to be knowledgeable about issues and project
activities because you must actively do something with what you see
and hear. If you passively look over a project, you will be unlikely to
absorb as much information. When you measure the work in a
project, you will define what information to collect, gather the information, and analyze it. In this process, you will understand what is
going on.

Effective measurement has many benefits. First, you are credible
to management and staff because people will respect your analytical
skills. Second, if you are the person doing the measurement, you gain
some political clout. Also, measuring will assist you in evaluating
different projects. It may clarify which ones are winners and which
should be redirected or terminated.

PURPOSE AND SCOPE

The objectives of this chapter are to address the following questions
for each major activity within the project life cycle:

- What are specific measurement goals?
- What is to be measured?
- How is the data to be collected and analyzed?
- How are the results to be interpreted and what are alternative decisions?

The scope of this chapter goes from the initial concept of the project to the completed project, since you will want to measure everything in between.

APPROACH

Here are the major items and milestones in the overall project that you should address during measurement:

- **The project concept**   Is the project concept sound and feasible?
- **The initial project plan**   Is the plan complete and accurate?
- **An updated project plan**   Does the plan reflect reality?
- **Money, work, and resources consumed by the project**   Are you on target?
- **The end products of the project**   Will you get what you paid for?
- **The project process**   Do you have an effective way to manage the project?
- **The project leader**   How effective is the project leader?

Assess the Project Concept

Goals of Measurement

A project concept is an idea about a project. It includes the purpose, scope, major milestones, and some idea of relative costs and benefits. How do you measure something this fuzzy? One goal is to assess if the project fits with the business goals and major thrusts of the company. For example, a project in an oil company had as its purpose the centralization of accounting. At the same time, management preferred decentralization. Because of lack of measurement, the project went on for six months before it was killed.

Another goal is to determine if the project is feasible in your company with your resources. Does the concept require you to be experienced in technology, products, or some area in which you have little experience? This is a key question when the project involves a company moving into a new line of business.

A third goal is to assess the impact of the project on other projects. If you fund this project, what resources will you take from other projects? Also, is this project necessary because of the requirements of other projects?

How to Conduct Measurement

If the desired end products of the project were in place, what benefits would your organization reap? How would business processes, the organization, and other factors be different and improved? Would your competitive position change?

If the concept passes this test, answer the following questions:

- Are the goals of the project clear and consistent with the focus of the company?
- Is the scope of the project concept compatible with the objectives?
- What things outside of the project concept are going to be required to support the project?
- If this is such a good idea, why hasn't it been thought of before? What is unique about this day and time?
- Has the company carried out similar projects in the past?
- Are the potential resource requirements in line with experience?
- Where is the inherent risk in the project?
- If you put together parts of the project concept, is the whole greater than the sum of the individual concepts? Are there any synergies between the parts?

The ideal world is to have all project concepts reviewed in a consistent way. The fact of life in many organizations is that projects are often started ad hoc or under the table. The idea some have is to

"Get it started and then after we show something, management will put money into it." This approach ensures that a lot of money is consumed in small projects that lack strategic benefit.

At the other extreme is the large project approach. Here management sets priorities and approves all project ideas. While this allows some strategic projects to start, it stifles useful projects that have a more narrow scope. Between these two extremes is where you want to be. A management steering committee should approve both small and large projects.

Alternative Decisions

If you decide not to proceed with a project as originally planned, several alternatives exist. A project concept can be shelved for a year. In most cases, this is the kiss of death of the concept. The concept can be approved for more in-depth work to generate a plan. Or a project can be combined with another existing or future project. Additional options are to change the purpose and scope and proceed, or fold the project into an existing project by expanding its scope.

Assess the Initial Project Plan

Goals of Measurement

Consider how to assess multiple projects. Let's assume that you have evaluated each plan and narrowed the field by eliminating those with low priority, high risk, and poor feasibility. What you are left with are good, viable candidates.

Your goal now is to select the mix of projects that can support both intermediate and longer-term objectives of the company. You also want to select projects that can stand alone. Otherwise, you may end up in a situation where all of the projects depend on a specific set of tasks in one project. You don't have to avoid risk in all projects. Consider having a high risk project every year as a means of having the organization extend itself.

How to Conduct Measurement

Once your initial plan is in place, answer the following questions to help you evaluate the plan.

- What is the contribution of the project to the company in the next year?
- What is the contribution of the project to the company in the long-term future?
- What risk does the project bring to the organization?
- How would you rate the fit of the project with the organization?
- What is the availability of staff? How good are staff members' skills?
- What technology risk does the project carry? What is the degree of technical complexity of the project?
- Is contractor/consultant risk involved?
- What is the size of the project? What resources are required? What is the cost?
- How much revenue will be generated?
- Does the project allow for competitive position enhancement?
- Does the project have leverage to support other projects and work?
- What is the available potential project leadership?
- What degree of interdependence with other projects does this project have?

Employing a formal method to evaluate projects forces management to be more aware and involved in the projects. One method is to evaluate the projects from the following perspectives:

- Corporate
- Business unit
- Systems and technology
- Risk
- Project leader and team
- Relationship with other projects

A common mistake is to apply inconsistent rules for funding projects based on politics. Another mistake is to base the allocation on fairness. When you attempt to be fair in project allocation, you give everyone a project or a role in a project. This tends to spread out resources. It is better to allocate resources to fewer projects where you will have major benefits than to pursue many small projects.

Alternative Decisions

These are basically the same options as for the project concept. Since you have gotten this far and more work has been performed, political fallout must be addressed. Should you tell members of a losing organization that their project has been deferred, or do you admit that it was killed off? Don't pursue either approach. Instead, visit the department and determine what the need was based on. See how this can be addressed without a formal project. For example, maybe the work can be addressed within the line organization.

Assess an Updated Project Plan

Goals of Measurement

Don't assume that the updated plan reflects reality—even if people say that they have been updating the schedule each week. The goal here is to answer the question, "Does the current project schedule and plan reflect reality?"

How to Conduct Measurement

Get a copy of the updated plan and the original approved project plan. Put them side by side. What is different? Where? Does the updated plan compare actual with planned schedules?

The updated plan in many cases will change, reflecting some or all of the following:

- Additional tasks appear due to unplanned work and rework.

- The updated plan becomes more detailed to reflect actual work.

- Resource assignments change as they mirror what really happened in the project.

However, you may not see any of these things. The resources may not be different; the schedule may not be different. This can mean that the plan is not being maintained or updated.

After the review, visit the person who developed the updated schedule and pose the following questions:

- How often do you update the schedule?

   The ideal answer is weekly or twice a week. Less than that leaves too much data that is not accurate. More detail than that means too much time spent in updates.

- How do you receive input about the schedule?

   Answers could vary from "I place a few calls" to "I go out and see the work."

- Do you verify what you are told? Or, do you accept what you are given?

- How do you fill in more detailed tasks? By yourself or with others?

- Who else reviews and uses the schedule? Is there any separate validation?

   Go out into the project and see where the work really is. Match the tasks you see being performed to the tasks on the GANTT chart print out. This will provide a healthy dose of reality. It may cause you to get rid of the existing plan.

Verify everything before going into budget vs. actual or projected work quality. If the people who are doing the scheduling see you doing this, they will be more careful in the next update.

Alternative Decisions

What do you do if the updated plan is not based on reality? You don't have the time or luxury to go back to the start of the project and recreate history. Begin with the tasks that are active and build a new set of detailed tasks. Next, go to the people doing the work and have them provide you with detailed tasks for the next month. Third, go back and review future estimates based on the plan and the first two steps. You now have a decent starting point for the actual results.

Before turning to the work of the project, let's consider project costs and measurement. Costs are based on work or hours and effort. Any analysis that you do for costs should first be based on work, as that is a more tangible item to measure than hours and effort.

Assess the Money and Resources Consumed by the Project

Earned Value Analysis

Earned value is based on the original estimate of the schedule and the progress to date and is used to determine if you are on budget. Let's take one task and work through a simple example.

The task characteristics are as follows:

- Planned or baseline duration—10 days
- Planned start date—May 1
- Planned end date—May 12
- Resources assigned: Person 1—$20/hour; Person 2—$10/hour
- Calendar—five days a week, eight hours per day

The total planned cost of the task is $2,400 (80 hours x ($20+ $10)). The actual start date is May 1. At the end of one week (May 5), the project is only 40 percent complete. Both resources have worked for the entire week. The actual cost incurred so far is $1,200. However, since you still have 60 percent of the work to go, your new estimate of completion is May 17, assuming that they are working along at the same rate and there is no new slippage. This is a slip of 5 calendar days and 3-1/2 working days and an overrun of $600. The new total estimated cost is $3,000.

Let's use this simple example to examine some of the concepts associated with earned value. Note that if you assigned no resources to a task, you have no work and no costs or earned value—another reason to assign resources to tasks.

- **Budgeted Cost of Work Scheduled (BCWS)** is the earned value of the task based on the plan. The planned percentage complete on May 5 is 50 percent. Multiply this percentage by the planned cost ($2,400) and you get $1,200 as the BCWS.

- **Budgeted Cost of Work Performed (BCWP)** is computed by multiplying the percentage complete (40 percent) by the planned or baseline cost ($2,400). BCWP for this simple example is $960.

- To find the **Scheduled Variance (SV)**, subtract the BCWP from the BCWS. Using the above figures, the SV is $240. If this were negative, you would be doing well since you would be beating your plan. Here you are not. You are in the hole.

- **Actual Cost of Work Performed (ACWP)** is the sum of all actual costs incurred to date. This includes any fixed costs incurred for the task. For you, this is $1,200. If you were to have the people work overtime, the costs would be higher.

- **Earned Value Cost Variance** is the difference between what was planned (BCWS) and what occurred (ACWP). This tells you if the task is on target with reference to costs. In the example the number is zero so you are on target.

- Total projected cost for the task is $3,000. This is also called **Forecast at Completion.**

- The **Cost Variance** is the difference between the baseline or planned cost and the actual cost. The variance is $600 and positive—you are going to overrun. If this were negative, you would underrun.

People like to use earned value to measure the budget or planned vs. actual, since this reflects cost as well as budget.

Activity-Based Costing

In activity-based costing, divide the project and organization into activities. With each activity, associate an average cost. The total cost of the project or product is the total of all average costs associated with all of the activities. Based on experience you can update the average cost of future activities. You can also compare actual average costs with planned average costs. Idle time is reflected in the difference between resource use and resource spending. This method is often used for scheduling and planning. The limitation of this approach is that it assumes stability in the activities. Activities and tasks can change.

Value Engineering and Kaizen Costing

In value engineering you want to reduce costs associated with work. This is target costing. You examine how to reduce the cost of each task. Kaizen costing is the continuous effort to reduce and control costs. In both of these you will focus on costs. These are equivalent to considering the same approaches relative to task structure and how the work is to be performed. One problem with this approach is that attempting to reduce costs in the middle of the project can actually drive costs up, due to the disturbance you cause. If you use this method, focus on tasks in the near future to minimize disruption.

Here are some guidelines:

- First, consider if the work is complete. If it is not complete, assessing quality is not meaningful. Have some of the tasks been pushed off to another part of the project in order to meet deadlines?
- In analyzing quality of work, consider the extreme areas. Look at the results of work for the most complex tasks. Then look at the results of the work done on simple tasks.

Assess the End Products of the Project

Goals of Measurement

When looking across multiple projects, the objective is to determine whether the limited attention and resources that can be directed toward reviewing milestones are pointed in the right direction. Ten active projects, each with ten milestones over a period of a year, would require you to review two milestones every week. This is too much for management attention.

After you decide that you are considering the right milestones, the next issue is to develop a consistent approach for all milestones to reduce overhead and to ensure consistency. Strive to be both efficient and effective, given the limited time.

A third goal is to ensure that the effects of the review are fed back to the team and appropriate managers and staff in an unfiltered and accurate way. Much disinformation may crop up after a review. Related to this is follow up on what is done after the review.

How to Conduct Measurement

Don't get bogged down with detail. Begin at a high level and determine the major milestones associated with each project. Create a project plan that contains each of the major milestones of the projects and summary tasks. You can now begin to pick and choose among the milestones. You can also assess when the reviews should occur. At the same time, you will be defining a milestone review process to be applied consistently across all projects.

The review process will identify the following:

- What documents, narrative, presentation, or proof will be offered for the milestone

- How the review will be conducted

- Who will conduct the review

- What the potential decisions regarding the milestone can be

- How follow-up and the decision on the milestone will be undertaken after the review

Experience shows that the more this review can be planned and orchestrated in advance, the better the efficiency and flexibility in coping with schedule changes. The review must be conducted consistently across all projects. Keep in mind the politics of the projects. Technology projects receive far more scrutiny than non-technology projects. This is because they are more interesting or unique, not because of the inherent risk. Don't stretch out a review. Apply whatever resources are required once it starts. Once a review is completed, issue the decision as soon as possible. Don't let the project team and others hang in the wind. That costs money and effectiveness. After a decision is announced, a follow-up process should be put into place immediately to ensure that the steps will be taken.

Alternative Decisions

You can accept the milestone as it is. You can reject it totally, though this is unlikely, since it is assumed that you have been involved in the process up to this point and had the opportunity to head off serious trouble. You can request rework and changes. These may hold up some of the project tasks. Finally, you may insist on changes subject to work being performed later. If you select the last alternative, ensure that you actively pursue this later. If you let it drop, you lose your credibility.

Evaluate the Project Process

Goals of Measurement

Here you want to evaluate the entire project management process. You want to answer fundamental questions such as: "Should I continue with project management?" "Can I improve how we do project management?" These questions deal with the fundamental issue of whether the effort of project management is worth the cost.

How to Conduct Measurement

Here are some questions you typically ask about project management:

- What is the quality of project managers? Are you getting better project managers? Do people want to become project managers?
- Are you getting good project ideas and concepts?
- Did the approved projects work out? If not, why not? If so, why?
- Should you have selected different projects at the start?
- What surprises occurred that you could have predicted, in hindsight?

- Which projects that were approved should have been placed on hold? Why?

- Are people learning from the projects and improving their skills and techniques?

- What was the greatest failure? What was the greatest success? Why?

- Are lessons learned gathered, interpreted, and folded back into the projects?

Create a table of how you are doing concerning the project process. Here is an example. The first column contains the phases of a project. The second contains your rating on a scale of 1 to 5 (1 is low; 5 is high). The third column indicates what you learned in this phase.

Phase	Score	Lessons Learned
Project concept		
Project plan		
Project approval		
Updating the project		
Cost and work analysis		
Issue management		
End products		
Project manager		

This table serves as a scorecard that clearly sets forth both positive and negative results. It is a key to showing management how the project management process can be improved.

Alternative Decisions

You can continue the process as is. This is unlikely, given that improvements are always possible. One common remedy for problems is to improve specific rows in the table. Another is to begin to treat projects with different characteristics differently. You can make the process more formal and bureaucratic. Alternatively, you can lighten up on part of the process. You can mix formal and informal methods. Make the approval of the plan and milestones more formal. This puts the spotlight on them. Other parts can be treated less formally.

Assess the Project Leader

Goals of Measurement

You are reviewing all project managers. You will find some good ones and some that require improvement. You can employ the comments and ideas about project managers mentioned earlier. Your goal is not to eliminate the poor managers but to improve their work.

How to Conduct Measurement

Begin by writing down the projects managed by each person over the past two years. Then assess each manager individually in terms of growth and improvement. Is the manager getting better at dealing with issues? How is the manager's ability to estimate and control resources?

Next, move up to project managers as a group. Rank their abilities in terms of problem-solving abilities, motivation and human resources skills, ability to control and direct the project, ability to cope with crises, etc. These topics could be columns in another table. The rows are the names of the project managers. Rate each manager on each attribute.

Alternative Decisions

Ask the best project managers that you identified to share their lessons learned and techniques with other project managers. Consider a mentoring program in which the best project managers spend time with the other project managers and guide them through issues. Try to salvage and improve a poor project leader rather than removing the leader.

EXAMPLES

The Railroad Example

The concept of mentoring was followed by most railroad lines. This was viewed as an inexpensive and feasible approach, given the distances involved and the time pressure to get the work done. When a new adjunct to a line was proposed (project concept), management considered the competition and potential revenue. Management also factored in the cumulative rail traffic from all of their projects. If the project was approved, a project manager was appointed to survey the route. Additional project managers were appointed for construction and logistics. Project managers were selected on the basis of their prior track record. Railroaders have always been great ones for telling and relating stories and experiences with each other. This was a prime method for sharing information and disseminating lessons learned.

Modern Examples

In the manufacturing firm, one of the biggest headaches was obtaining the status of the project when work was being performed at remote sites on the network. Calls and electronic mail to five people would get five different answers. This meant more calls. Another problem was determining when unanticipated work or extra work was being performed. This was only discovered when the bills came in from the contractors or the timesheets were turned in by employees. Payroll and accounts payable had to be contacted to block processing. Vendors were placed on notice that there would no payment for additional, out-of-scope work without the project leader's prior approval.

In the construction firm, problems in measurement centered on trying to measure how much progress was being made in establishing a standard project framework. What constituted success and completion of tasks? The decision was to report the work complete only when the managers were employing the standard template in their work.

GUIDELINES

- **Match up the milestones in the plan with the original goals of the project.**

  Projects with unrealistic goals are often reflected in unreal milestones. Milestones seldom deliver more than the objectives; it is not uncommon for the milestones to deliver less.

- **If management asks for frequent reviews, target small tasks that can be done in a day.**

  Measurement and review of a project can be excessive if a project is receiving too much management attention. Some projects seem to be in a state of almost continuous audit. Management wants daily reports. Target small tasks that can be done in a day to show progress.

- **Enforce policies or they will have no meaning.**

  Policies are rules that the project manager and team are supposed to follow. Put teeth in the policies in terms of review and enforcement so that people pay attention to the policy. A caveat here—if you enforce the rules, then you have to be flexible in interpreting the policies for each project. A policy that fits a large project can swamp a small project with overhead.

- **For good project quality, focus on good integration of project parts.**

  Quality of individual parts is relatively easy to determine. Quality of an integrated subsystem is more difficult to assess. Look for integration problems and then trace these back to components in the work.

- **Provide interpretation of the goals from different perspectives—management, the project, line organizations, the company as a whole, etc.**

Whether or not a goal is achievable or has been achieved depends on point of view. The project concept and plan define specific goals and targets. Even if clearly stated, there is room for interpretation based on the audience. It is best and safest to look at the issue from different points of view.

- **Use measurement to keep credibility for the plan.**

Lack of measuring a project can result in lost credibility for the plan. Without measurement you really don't know what state the project is in. You will likely be surprised when a major milestone slips or is unacceptable. Then it may be too late to recover.

- **Conduct post-project reviews to capture lessons learned.**

Gather lessons learned as you work on the project. However, you are also under the constraint of time. Often, the best time to define lessons learned is after a major phase of the project and at the end of the project. The experience is still fresh in your mind, yet it was not absorbed in daily work.

- **Make external comparisons with other projects.**

Comparisons with other projects are often useful if put into place early. Putting a standard comparison method in place at the start of projects makes life easier. People will then expect and anticipate the comparison. They will understand how they are being measured and respond positively.

STATUS CHECK

- Do you have a formal process for moving from a project concept to the project plan? Or do plans just get created? How many of the plans created actually get started?

- In retrospect, of the plan ideas that were shelved, which would have been winners and should have been approved? Go back and review why they were killed.

- To what extent are plans reviewed in terms of accurately reflecting the status of the project? Are costs and analysis developed based on the assumption of accuracy? How often have cost and work problems surfaced that were later proven to be false when people checked the actual state of the project?

ACTION ITEMS

1. First assess how your project is doing, using the measurement methods of this chapter. Then you can expand this method to multiple projects.

2. Define to what extent your organization addresses each of the project management areas on a formal or informal basis. What changes would you suggest? What benefits would accrue from your changes?

CHAPTER 15

MANAGING PROJECT ISSUES

CONTENTS

15

MANAGING PROJECT ISSUES

INTRODUCTION

In the past, many people looked at issue management as a topic impossible to generalize about, since each project was viewed as different. Issue management was given little attention. However, managing issues is a key ingredient to project success. An issue is something that must be addressed; otherwise, progress may slow and the project may deteriorate. An issue can be a problem or an opportunity. It can relate to something within the team or to technical, managerial, and political problems. How and when issues are handled impacts the project schedule and the plan. If not addressed, an issue can blossom into a full-fledged crisis. In this chapter we will discuss how to prevent an issue from becoming a crisis. Chapters 18 and 19 are devoted to handling a crisis.

At any time in a project, active issues may or may not be interdependent. Most project problems and slippage can be traced to specific issues. Issue management tests the range of a project manager's capabilities far more than project control or project administration. Required skills include identifying an issue, collecting data, performing analysis, developing alternatives for resolution, obtaining concurrence on the solution, selling the solution, and implementing the solution.

PURPOSE AND SCOPE

The purpose of this chapter is to help you manage and direct the outcome of single and multiple issues. You will be provided with guidance on identification, analysis, decision-making, and implementation of solutions. The scope includes all of these activities, across the entire project.

A systematic approach is more efficient in addressing issues. Through analysis of multiple issues, you will be able to address families or sets of issues. By tracking issues you will add to your lessons learned.

APPROACH

Let's examine a list of issues that you might encounter in a substantial project. The list below has been drawn from projects tackled over the years. The potential impact of each issue is indicated.

List of Sample Issues

ISSUES	POTENTIAL PROBLEMS
Project restructuring	Parallel effort and reduced project time
Loss of key a person from the project	Slowing of the schedule
Team morale drop	Reduced productivity
Line manager opposition	Road blocks to decisions
Falling behind schedule	Milestones slippage
Expanding scope	Milestones slippage
Competition for money and resources	Slowing of the schedule
New regulatory changes	Changes in nature of the project work
Conflict over work assignments	Lost productivity and low morale

HOW TO SPOT POTENTIAL PROBLEMS

These questions apply to opportunities as well as to problems. For each question, rate an issue on a positive or negative scale from −3 to +4, where positive numbers indicate benefits and negative numbers are disadvantages. The rating 0 means no impact. When finished, you can generate a bar chart such as that in Figure 15.1. Two examples for the manufacturing company are given in Figures 15.2 and 15.3.

- What is the urgency of the issue to the project?
- Are other organizations affected by the issue?
- Which business processes are impacted by the issue? To what extent? Include here procedures and policies for the process.
- How does the issue relate to the systems and technology in place?
- Does the issue have any impact on the company overall? For example, are many side effects generated by a decision on the issue?

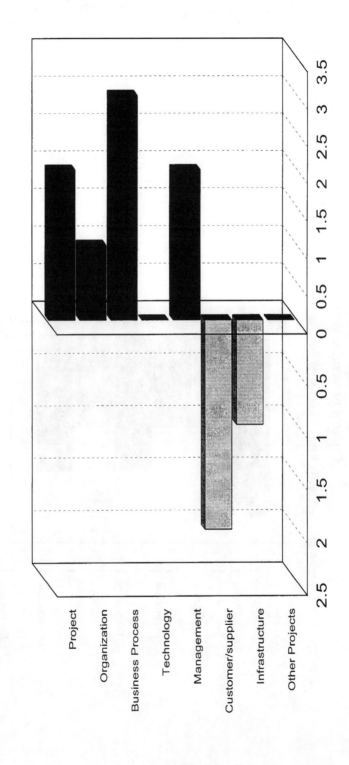

Figure 15.1: Sample Rating of Issue

Figure 15.2: Radon Control of Vendor Work

In this example, the issue has a negative effect. The project, infrastructure, and technology are affected the least.

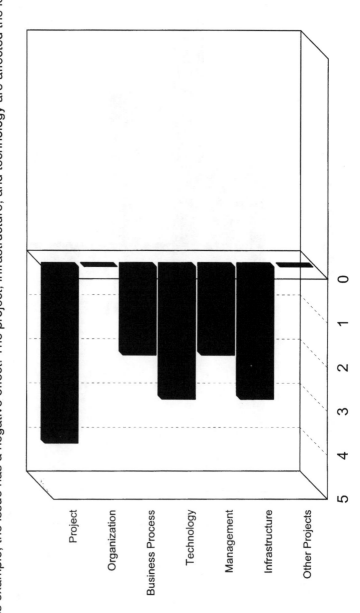

Figure 15.3: Radon
Potential Faster Network Devices

In this example, there are clear benefits to faster local area network links.
However, acquisition of the technology could impact the project schedule.

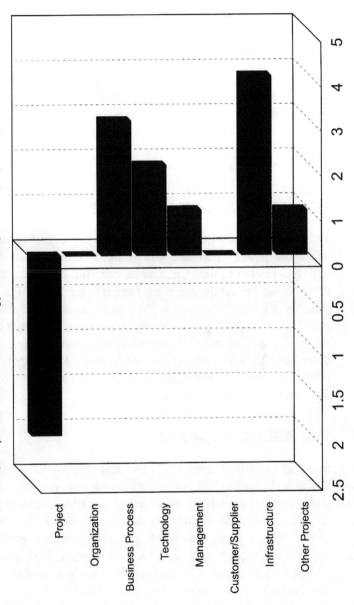

- Is there any fallout from the issue on the customers and suppliers?

- What is the impact of the issue on the infrastructure (buildings, office layout, parking, telephones, etc.)?

- What is the effect of the issue on other projects?

Determine to leave the low priority issues alone unless they are grouped with a high priority issue. At the construction firm, it was decided that work would be performed only on issues with ratings of –2, –3, and +3. These are the most severe issues in terms of either benefits or costs. The other issues were put on hold until they grew more beneficial or were grouped with an important issue.

HOW AN ISSUE BECOMES A CRISIS

When an issue surfaces, you are faced with several choices. If you adopt the wrong approach, you could make the issue worse. Just calling attention to the issue can be harmful. For example, suppose that you notice that the project is falling behind schedule. You could react by telling everyone that "We must all do more work to catch up." This may instill panic. People on the team may react by slowing down. Another reaction is to recruit a number of new team members. They have to be brought up to speed. This takes people away from productive work. Coordination and decision-making are slowed down. The issue grows into a crisis.

THE ISSUES DATABASE

Seek an organized approach when tracking issues. A database is useful and requires minimal effort. In the chapters on methods and tools and the Internet, database management systems and groupware are discussed. Both can support an issues database. Another alternative is to use paper. However, generating the reports and doing analysis will require more intensive manual work. With many projects, all three of the modern examples saw the wisdom of establishing a standardized available system on a network.

Having each project manager spend time developing databases that are not compatible with other project managers' work is not the best use of time and effort.

Here is a list of data elements to employ:

- **Identifier of the issue**   Use a separate code for each issue.
- **Status of the issue**   Based on where the issue is in the life cycle, sample codes might include the following:
  I—identified
  A—assigned and being analyzed
  AD—awaiting decision and resolution after analysis
  R—resolved
  F—followed up on
  RE—replaced by another issue
  T— terminated or eliminated
- **Priority level of the issue**   Define several levels of priority:
  A—extremely important in that the project is impacted within days if not resolved
  B—the project will be impacted in weeks if not resolved
  C—impact is marginal on the organization and project
- **Organization impacted**   This is the major organization affected by the issue.
- **Date the issue was created**   This is the date the issue formally begins to be tracked.
- **Description**   This is a summary description of the issue.
- **Impact**   This field identifies the effects of not addressing the issue.
- **Related tasks**   These are the tasks (by number) in the schedule that are impacted by the issue.
- **Related issues**   This includes how the listed issues are related.
- **Person assigned to the issue**
- **Date of expected resolution**
- **Resolution code**   Examples:
  R—replaced by another issue
  D—decided
  S—shelved indefinitely
  T—terminated

- **Decision on the issue**   This is a statement of the decision made.

- **Actions**   These are the actions that flowed from the decisions made.

- **Comments**   This field is for free-form comments on the issue.

Typically, each issue is associated with a series of events or actions. An event log would have the following elements:

- Identifier of the issue
- Event number—A unique number assigned to the event
- Event date
- Person recording the event—May be different from the person responsible
- Type of event—Meeting, telephone call, fax, and e-mail
- Result of the event
- Comments

The event log links to the issues database using the identifier of the issue. The index to ensure uniqueness is a combination of the identifier and the event number. How would you use these files? Set these up as databases on a file server. The information can be accessed by the project team members. Access to update the log could be controlled.

Use the database to summarize issues for a project. A sample rating is in Figure 15.4.

Across multiple projects you can integrate information by the following criteria:

- **Priority of the issue**   This can isolate all high-priority issues so that management can address the entire set of issues on an organized basis.

- **Organization**   This can indicate the extent to which issues are impacting specific organizations.

- **Families of issues**   By clustering by families of related issues, you can attempt to deal with groups of issues as opposed to single issues.

Figure 15.4: Example of Issues Report for One Project

Priority: High

Type of Issue	Date Opened	Issue ID	Issue	Status Closed	Date	Resolution
Work	3/1/97	005	Ability to track work performed remotely.	Open		
Technology	4/1/97	008	Decide on network card	Closed	5/15/97	100 Mbps Ether net card

This table is from the Radon example. It assumes that the person who receives the report is familiar with the issue so that a description is not necessary. The report is first sorted by priority. The other sorts are by type of issue and date opened.

You can also analyze the issues by aging analysis. That is, you can develop a histogram of open issues over time. Consider Figure 15.5. There are lines for the total number of issues for all projects and for the number of high-priority issues.

COMMON MISTAKES IN ADDRESSING ISSUES

Over time, we have identified eight recurring problems or failures in addressing issues. Let's discuss each of these in terms of how it can happen and the impact:

- **Failure 1: Being unaware of the issue**

  If you keep too narrow a focus, you will find that you are missing details. You will be missing signs and symptoms of problems. Instead, be constantly on the lookout for more issues.

- **Failure 2: Misdiagnosing the issue**

  Once you have identified the issue, a common error is to plunge in and attempt to address the issue without analysis. Misdiagnosis of the issue is likely. Then you either make the issue worse or lose credibility as a manager.

- **Failure 3: Not selling the decision to management**

  After the issue is identified and the analysis performed, a decision is made and seems logical. If the project manager jumps in to act on the decision without selling it to management, problems may occur. Actions taken as a result of the decision can affect resources, costs, and the schedule. If management wasn't consulted, the bill for these extra costs may be a shock. Failure to market the decision to management opens the door to attacks on the decision as well.

- **Failure 4: Making decisions without planned action**

  A decision is announced. Everyone who hears it asks, "What does it mean?" The answer is "nothing," unless the decision is followed by action. Some people seem to think that they can announce a decision and then wait for weeks to take action. As time passes, the credibility of the decision is questioned. When

Figure 15.5: Open Issue Graphs

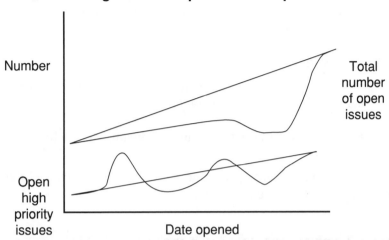

The purpose of these graphs is to draw attention to open issues that were opened long ago. These are the aged, high priority issues. The total number of open issues graph is typical in that overall it is the most recent issues that are unresolved. The other graph of open, high priority issues is troubling. It shows two humps of issues that are aged. This means that there are substantial old and high priority issues remaining.

the action finally comes, the situation may have changed, making the decision inappropriate.

- **Failure 5: Acting without the framework of a decision**

  Another problem occurs with project leaders who are action-oriented. They move from their assessment of the problem to immediate action. This is fine in a true emergency. However, this can be deadly. First, the actions will appear as chaotic without the framework of a decision. Second, the actions will probably be incomplete, requiring additional actions. These may contradict or overlap the previous actions.

- **Failure 6: Failing to act when you should**

  Some people cannot decide when to act. While many favor a conservative approach, action must be taken immediately after the decision is announced.

- **Failure 7: Acting when you should wait**

  This is a common mistake with new project managers. They make decisions and take action on the spot.

- **Failure 8: Taking actions that are inconsistent with decisions made**

  This occurs because people do not think through whether the actions support the decisions.

SEVEN STEPS IN MANAGING ISSUES

Step 1: Recognize the Issue

It often starts with a question or offhand comment. "What's happening with Harry?" is a question leading to a personnel issue. "I heard you won't need that piece of equipment by the first of the month after all." This can imply that the project is behind schedule. A verbal message may be the first symptom of an issue.

Respond by answering the following questions:

- **Does the symptom relate to a current active issue?** Is it just another symptom of a known problem? If so, employ this information to gain a better understanding of the current issue with which it is associated.

- **Can the symptom be grouped with anything else going on?** You can group by organization, technology, management, customer, and supplier. If you see no such connection, wait to raise the issue.

- **What are the characteristics of the issue?** At this point, define the issue. Use the database of issues and fill in the elements. A form is included in Figure 15.6.

Figure 15.6: Sample Issue Form

Issue Management Form

ID: _____ Name: _____ Priority: _____

Title: _____

Description: _____

_____

_____

Impacts if not addressed: _____

_____

Assigned to: _____ Date assigned: _____

Issue Activity

Status	*Date*	*Who Entered*	*Action/Result*

Date Resolved: _____ How Resolved: _____

Comments: _____

_____

- **What priority should be assigned to the issue?** Set priorities by urgency of the project. Do not use other criteria, such as benefits to the organization or management, since mixing criteria complicates decisions on priority.

- **What should be done with the issue initially?** Discuss it with the project team to collect ideas and to see who has the most interest in the issue. Assign the issue to someone who cares about it. Giving an issue to someone who dislikes the subject will result in it getting little attention.

Step 2: Analyze the Issue

Use a combination of direct observation, interviews, review of documents, and meetings to collect the information for the analysis. In collecting the information, don't draw attention to the issue. Instead, talk about symptoms and impact. If you zero in on the issue, people may expect too much in terms of resolution. Also, by tagging the issue too early, everyone accepts the preliminary definition, which may be in error.

Start with the person who proposed the improvement and collect as much information as you can. Find out how the project team and the work can be affected, as well as the end products of the project.

Categorize the issue for the database, using the topics and information earlier in this chapter. Next, draw up the following table. This table allows for different interpretations of the issue (from conservative or minimal to radical). Obtain different views from the various members of the project team and others. For each interpretation, list the symptoms of the issue in one column, the impact of the issue in another column, and the principal dimensions of the issue, based on the interpretation, in the last column. An example for the manufacturing company is shown in Figure 15.7.

Interpretation	Symptoms	Impact	Dimensions

Figure 15.7: Example of Categorization of Issue for Radon Control of Vendor Work

Interpretation	Symptoms	Impact	Dimensions
Lack of control over vendor	Excessive invoices; invoices for unapproved work	Cost overruns; schedule slippage	Accounting and financial controls
Weak project management	Lack of direction to vendor	Lose control of project	Project management
Weak central management	This is only one example of weak controls	Decentralized organization undermines any central initiatives	General management

Now focus on the effects and benefits of the issue on the project itself. Construct another bar chart, using the following categories. An example is given in Figure 15.8.

You can also use the following perspectives to analyze the project:

- Project team
- Tasks and work performed
- Methods and tools employed in the project
- Project management and control
- Schedule of the project
- Costs and resources required
- Quality and nature of end products and milestones
- Interfacing projects

Step 3: Define Alternatives

As you perform an analysis, also define alternatives for decisions. Consider the following suggestions or actions and estimate the effects that will accrue to the project, organization, or team.

Alternative 1: Do nothing.

This is the alternative to adopt most often. Wait and let the issue mature. As it does, the impact of the issue will become more evident. Note that even if you do not take action, you continue to track the issue; you still treat the issue as valid.

Alternative 2: Restructure the project with no new resources.

This alternative helps you see what you can do with the resources you have. It forces creative thinking in organizing the work.

Alternative 3: Apply resources to the issue without regard to cost.

If you consider applying resources to the issue in virtually unlimited amounts, you can see the limits of what can be bought. This is an important alternative because it reveals the true limitations of resources. When you apply additional resources, you actually slow the project, since there is more

Figure 15.8: Radon Example
Distribute Project Authority to Regions

In this chart, there are potential negative effects for the work, commonality of tools, and more complex project management. On the other hand, the schedule might improve. Quality with local oversight will improve. Other projects are not impacted. The cost and resources might be less using local resources

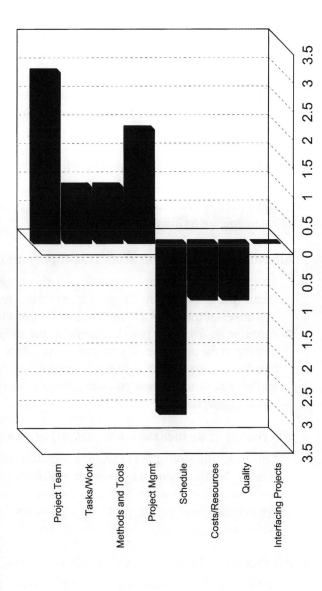

coordination involved in the handling of the resources (bringing people on board and setting up equipment, for example). Always be ready to answer the question, "What could you do with additional resources?"

Alternative 4: Reassign resources within the project team.

This alternative is often useful for personnel problems and conflicts within the project team. Consider setting up subproject teams of different people to see if that works. This alternative tests your knowledge of the project team and also defines the limits of their flexibility.

Alternative 5: Remove resources from the project.

This could be applied to personnel issues. It could also be considered after a major phase of work in the project is finished, when you are trying to downsize.

Alternative 6: Expand the scope and/or purpose of the project.

The issue may be very important to the project. However, taking the issue on in the project will lead to the scope or objectives of the project being expanded. If this is a possibility, consider an overall strategic change in the project to address more than the one issue. If you let the scope expand naturally to handle the issue, you are likely to run into schedule problems and resource shortages as you scramble to address the expanded scope or purpose.

Alternative 7: Reduce the scope and/or purpose of the project.

This is the flip side of the previous alternative. If you encounter a major obstacle, you may wish to avoid direct confrontation with the cause. Instead, you might consider downsizing the project and avoiding the problem. This is not cowardly, since with success you can often later return and expand the scope.

Alternative 8: Treat the issue outside of the project.

Under this alternative, you are attempting to insulate the project from the issue. You will attempt to address it away from the project.

Alternative 9: Change the mix of methods and tools in the project.

Look for a better way to do the work, using better or different methods and tools. The trade-off against the potential benefit is that there is a learning curve for the new approaches.

Step 4: Make Decisions

After considering alternatives, think through each in terms of defining the potential actions that will flow in support of the decision. Employ the following table to analyze the situation further before making a decision.

1. Issues Addressed	2. Alternative Decisions	3. Actions	4. Effects	5. Risks

Select different issues to address (column 1). For each, identify several alternative decisions (column 2). For each decision, identify actions that will implement the decisions (column 3). What would be the likely effects of the actions on the issues? This is column 4. These effects are not guaranteed, which leads to column 5, in which risk is identified. This approach will allow you to resolve incidents of conflict in resources.

The actions typically involve politics, changes in the plan, and resource actions. After you select the decision and identify the actions, inform management of the issue and the recommended approach.

Step 5: Announce the Decision and Actions

Announce the decisions and the actions at the same time. Complete paperwork on the actions prior to the announcement. This strategy will let people know that you are serious. Also, benefits will begin earlier and people will be more likely to be supportive.

Step 6: Take Action

Take all actions at the same time. Change the project team, resources, plan, and scope all at once. If you attempt to do it sequentially, you will end up having a mixture of old and new for some period of time. You want to establish an entirely new mindset. This is another reason for considering bundles or groups of issues at one time.

Step 7: Follow Up

The results of the actions and decisions should be seen quickly after implementation for most issues. In follow up, ask several questions:

- Did the issue get addressed or were only symptoms treated?

- Are side effects of the actions and decisions creating new issues?

- Do additional areas exist in which one can apply the same actions and decisions to handle even more issues with little more effort?

MANAGING GROUPS OF ISSUES

Typically, you have to address many issues. They are in different stages of being identified and addressed. How can you more effectively manage groups of issues?

Here are some guidelines.

- View all of the open issues at one time and compare the mix of issues now with that of months ago. Do you see some lingering issues? Has the number of important issues increased? Has the average time to resolve an issue increased?

- If you devoted a solid day to working on issues, what could you accomplish? What things get in the way of resolving the issues?

- Do you know the status of the issues? How effective is your tracking?

EXAMPLES

The Railroad Example

Building the railroads was a constant series of battles over issues. Political issues concerned obtaining support from the government. Technical issues related to how specific barriers would be addressed. The railroads presented a constant logistics challenge in which men, materials, and equipment had to be obtained, staged, transported, set up, maintained, and then moved again. Most railroad administrators divided the issues into logistics, engineering, marketing, and finance. To link issues and keep track of what could be done across families, management had to track the issues manually.

Issues that surfaced on the road or in the field were addressed by the managers at the remote locations. The escalation process was through the line organization, augmented by direct contact from the field to the headquarters.

Modern Examples

The leader at the manufacturing firm thought that issues could be handled as they arose. This worked for the first ten or so issues. Then things became more complex. The elapsed time to resolve an issue dragged out and more issues were active at the same time. People would call with status reports on issues. The names and titles of the issues were very similar since they all involved the network. This created more confusion and return calls. To compound the situation, management wanted to have information on specific topics. Reviews of these requests indicated that they were tied to the issues.

After the issues database and formal approach to issues management were implemented, the situation improved somewhat. However, the project leader was the only one doing the entry and tracking. The problem was that the issues management function was not made into a collaborative effort with the field project leaders.

The construction firm had experience with issue tracking. The firm tracked by type of material, contractor, location, and other criteria. This became invaluable to the project. Since the project leaders already were oriented toward projects and issues, project management was made easier by relating to issue management.

A common organization culture base for issues management was lacking at the consumer products company. Product leaders had their own styles and approaches. To retrain all of these employees and complete the project proved impossible. So the project leader was forced to simplify the approach that was presented. He created a list of top ten issues and attempted to create a new culture and climate around these ten issues. He borrowed the idea from the criminal justice system. He was successful, but he never did change the overall climate. This would have required a much greater management commitment.

GUIDELINES

- **Deal with political problems in a project to avoid project collapse.**

  Political issues in a project can undermine morale. If several are allowed to fester and multiply, the situations will feed off of each other and get worse.

- **Consider large and small issues together.**

  Patience in a project includes not jumping on easy issues. It is a real temptation to resolve simpler issues quickly. Everyone feels good that something is getting accomplished. This is a false sense of security. If the bigger issues require resolution that undoes what you just announced, your credibility is lost and you have to start over. This is another reason to consider issues in groups.

- **Determine the relationship between issues.**

  A few underlying causes can generate many seemingly independent issues and many more symptoms. What can happen is that several issues or causes of issues join hands and begin to cause impact across the project. More symptoms are generated.

  Group issues by the areas of the project such as the resources used, the organization involved, or the method being used. Take action on a group of issues, if possible.

- **Reward people (including yourself) for identifying issues.**

  This motivates you and others to be on the watch for new issues. It can be overdone if people just complain. However, if the focus is on finding issues or suggesting improvements, the complaints have a positive effect.

- **Know which issues are constraints.**

  A constraint is something that you cannot change. What issues do you accept as constraints and which do you address? This is a key decision, since it determines how the issue is interpreted.

  Look at a stand-alone issue as a constraint. A stand-alone issue is rare and exists in a vacuum. No other issues are involved. Whatever action is taken will impact only that issue.

- **Decide whether an issue is secondary or artificial.**

  For every real issue, a project may have 5 to 10 artificial issues. When symptoms of problems appear, you may fix a symptom only to see another appear if you failed to pinpoint the real issue. Before you treat symptoms, decide if the issue represented is artificial or secondary.

- **Analyze the underlying source and concern behind the issue.**

  What is behind the issue is almost more important than the issue itself. The issue may be just one manifestation of an underlying problem. Understanding the cause of the issue can lead you to discover whether additional issues stem from the same cause.

- **Know when something is out of scope to keep a project on schedule.**

  Many issues relate to parts of the project that are not within the original scope of the project. When one of these comes up, the team and leader cannot deal with it quickly since it is beyond what was originally in the project. Test the scope with each issue.

- **Delegate the task of researching issues and follow up in tracking.**

It is impossible for you to follow up on every issue. You must delegate. To be successful, track how the follow-up is going and how the issue has changed or has been transformed through further analysis.

- **With issues involving outside vendors, prevent one vendor from criticizing another.**

  This is not an unusual practice. Be aware that it can happen and point this out to each vendor prior to a meeting. During the meeting, if a vendor is not present, try to mitigate any criticism of the vendor and turn the conversation into action steps regarding the issue.

- **Address basic project issues to prevent the project team from working on less important, though visible, work.**

  During a project, management issues can emerge. Usually no emergency exists for most issues. However, if the elapsed time without resolution continues to grow, the team may get the impression that management does not care or is not interested in the project. The team may think that the issue is not important. This can lead to a lack of confidence among team members. The team members may work on tasks that are not impacted by the issue. If the issue is important, the tasks related to the issue are usually important.

- **Test the degree of flexibility in a plan gradually rather than precipitously.**

  Changes in a project occur over time. Change tests flexibility. Change is often generated by the solution and resolution of issues, so change depends on how issues are sorted out. Each time you resolve an issue, you may change the plan, testing the flexibility.

- **Accompany changes in project leaders with a change in project scope.**

  When you make the major change of replacing a leader, consider other changes, including project scope.

- **Take time to gain an overall perspective after setbacks on issues.**

  You cannot expect to win on every issue. What do you do when you lose? Take an overall perspective on what just happened. You will gain a more general view of the situation and gather lessons learned for the future.

- **Resolve political issues to prevent polarization of a project team.**

  Political issues that remain unresolved can begin to split the project team. Work to resolve these issues and prevent the team from working against itself.

STATUS CHECK

- Assess your abilities in each of the following areas:
  —Identifying issues
  —Determining the ramifications and impact of issues
  —Deciding on the issues and timing of announcements
  —Defining the action items to be taken

- Evaluate your organization in dealing with issues. Answer the following questions:
  —Does the organization have a standardized issue management approach?
  —Does the organization have standardized software for issue management?
  —What escalation process is employed for issue resolution?
  —Are decisions clearly separated from actions? Are actions linked to decisions?

- Evaluate how well your organization tracks issues by answering the following questions:
  —Do you know how many issues are open and closed in a specific project?
  —Do you know which issues have the highest priority?
  —What is the age of the oldest outstanding issue?

ACTION ITEMS

1. For a project that you are working on, identify several issues (both open and closed). Determine the step that each is currently in. Next, assess how effectively each issue was addressed in previous steps.

2. Try applying the alternative decisions that were identified in the issues in Question 1. What did you learn about the issues after doing this that you did not know before?

3. Look back at several attempts to address issues. Were the benefits and effects as anticipated?

CHAPTER 16

MANAGING MULTIPLE PROJECTS

CONTENTS

16

MANAGING MULTIPLE PROJECTS

INTRODUCTION

In ancient times, the attention was on single, large projects. Such was the case with the pyramids and other Wonders of the World. Change began to occur with the Roman Empire. Given how far apart the cities were and the need for erecting the basic buildings and aqueducts, Rome implemented parallel building teams within each region of the Empire. Teams would construct the temple, forum, theater, and stadium. Then they would move on to the next town. Most of the construction was carried out according to standard plans and templates. However, in some cases the building materials were not suitable or the building was to be unique. A body of specialists emerged to deal with such situations. As a result, the Romans were able to cope with multiple projects.

Another example of multiple projects is found in the days of railroad construction. Almost all railroads had to implement multiple, parallel projects in order to meet schedules.

Today, you deal increasingly with multiple projects for the following reasons:

- People have learned to break up large projects into several smaller projects. Smaller projects are easier to manage, though coordination is required between the projects.

- Projects today potentially have many elements in common so that interfaces and coordination are a part of project management.

- Management wants to manage resources more efficiently and make decisions that apply to all projects.

To manage multiple projects, first consider how projects can be interdependent. Then consider how to direct multiple projects with different sets of goals. With these steps accomplished, you can turn to discussing the specifics of management.

Many of the benefits and results from project management occur in the management of multiple projects. This aspect of project management enables you to carry out trade-offs in resource use. You can gather lessons learned from one project and apply them to other projects. Managing multiple projects allows you to deal with issues and status across the projects. You can observe the impact of management style and practice on specific projects.

Interest in collaborative decision-making and project management has been increasing. In this environment, managers and staff work together to identify and resolve issues and conflicts. When properly employed, this approach is cost-effective and self-sustaining.

PURPOSE AND SCOPE

The purpose of this chapter is to help you monitor and manage several projects at the same time. These projects may have many elements in common, or they may share nothing. An overriding purpose is to manage multiple projects so that the projects in total are assisted and not harmed by management. You can benefit from suggestions on how to exploit synergy between projects. You also want to ensure that future projects benefit from lessons learned in past projects.

The scope of the effort includes all active projects that you have at one time as well as those in the future. Even if the projects are independent, an overall understanding can aid in the progress of the individual projects.

APPROACH

Managing multiple projects is growing in importance. Several changes have contributed to this trend. One is growth in the use of project management. The second is that of downsizing and rightsizing. A third is the increase in the use of technology.

Projects are competing more for the same scarce resources and for management attention. Coordination is mandated when multiple projects touch the same organizations, set of suppliers, or customer segment.

Fortunately, the technology is in place to support the management and coordination of multiple projects. In the days of mainframe computer dominance and stand-alone PCs, coordination of multiple projects was difficult. Networking, the Internet, and client-server computing all have made managing multiple projects easier and more affordable.

In a single project, few opportunities arise for trade-offs and decision-making unless the project is very large. You are often faced with yes or no decisions. Ideally, single projects have one project leader, one set of tools and methods, and a unified single purpose.

When you move up to multiple projects, the situation changes, opening up new positive possibilities, such as the following:

- Opportunities for resource trade-offs between projects

- Opportunities to set priorities and define policies across multiple projects

- Opportunities for greater economy due to methods and tools being deployed across more projects, people, and organizations

A negative aspect of multiple project management is the possibility that diversity will lead to misunderstandings, lack of communications, and other similar problems.

Some of the benefits of turning one large project into smaller projects are as follows:

- Greater parallel effort is achieved

- Coordination within the project is simpler, since the individual projects are smaller and less complex

- Less bureaucracy and more attention to issues are possible

- Problems and opportunities have greater visibility because they cross organizations

Potential risks are that you now have to manage multiple projects and that the potential exists for greater disparity between projects.

HOW PROJECTS INTERRELATE

When you consider multiple projects, reflect on how they can interrelate. What do they have in common and what do they share? Efficient management of multiple projects involves being able to group projects.
Here are some factors to consider in analysis and grouping:

- **Technology**

    The projects may share the same technology. This is the case with the manufacturing example. All of the organizations in the various countries share the same technology. Such projects compete for people with expertise. The payoff in managing this group of projects efficiently is the sharing of technology-related resources and the sharing of lessons learned.

- **Resources**

    Projects may be similar and compete for people, equipment, or facilities. Resource allocation among competing projects is a major focus of managing multiple projects. The benefit of managing resources well among multiple projects is the more efficient deployment and allocation of resources.

- **Time**

    For projects occurring in the same period or in overlapping periods of time, it is possible to share resources, project templates, methods and tools, and technology.

- **Project Templates**

    Projects not only may draw upon a common resource pool, but also may fit within a common task framework using the template that was discussed in Chapter 2.

- **Methods and Tools**

    Projects can employ the same methods and tools. The benefit is the opportunity to capitalize on lessons learned and to share expertise in the use of the methods and tools.

- **Direct Dependencies**

  Some projects are directly interdependent. That is, the milestones and activities in one project are employed by other projects. This is clearly a high priority grouping when you are considering an overall schedule.

- **Organization**

  Projects can be grouped by the organizations that they affect or involve. The benefit here is to center attention on a group of projects that impacts a single organization.

- **Management**

  You can group projects that fall under the same general manager. This is similar to, but not identical to, the organization approach above. It has appeal as a way to group projects because one manager at a high level can make decisions for the group.

- **Business Processes**

  Projects may be grouped according to which ones affect the same business processes. If these are key business processes (e.g., order entry, inventory, accounting), this grouping is very important. Otherwise, changes brought about by individual projects could disrupt the business process, affecting costs, revenues, and service.

- **Customers or Suppliers**

  Here you are grouping projects according to whether they relate to or impact the same customer or supplier segment.

Consider all of your projects using the above groupings, since these groupings will help you by providing alternative perspectives. Note that you can have overlap, since several projects may share more than one of the above list in common.

OBJECTIVES AND BENEFITS OF DIRECTING MULTIPLE PROJECTS

Here are some objectives to keep in mind as you manage several projects:

- Synergy among the projects
- Smooth interfaces and shared resources
- Effective and efficient use of resources
- The sharing of cumulative experience so that the lessons learned among several projects are passed on and employed by a future project

These are reasonable goals even if they are not easily achieved. Assuming that you reach these objectives, experience and observation show that the following benefits will accrue:

- In projects which have dependencies, the flow of information and interface between the projects is such that there is no misunderstanding or miscommunication. The schedules of the future projects are more likely to be met.
- Projects that have common resources in the same time frame are able to resolve resource conflicts through collaborative decision-making at lower levels of the organization, thereby reducing project management overhead and providing a clearer direction for the projects.
- Projects sharing the same methods and tools benefit from the shared experience and lessons learned so that productivity in the projects is better than if they were separate.
- Management has greater control and direction over projects. Using a collaborative project management approach, the project manager works with established frameworks and structure to achieve results. This extends to resolving conflicts, using the project template, gathering lessons learned, and resolving issues.

BARRIERS TO EFFECTIVE MANAGEMENT

It is evident that many benefits derive from actively managing and directing multiple projects as opposed to treating them as separate, individual projects. Why don't all organizations with several projects do this?

Observation has shown the following major constraints and barriers to multiple-project management:

- **Corporate culture**   In some companies the focus is on achieving short-term project goals. Gathering lessons learned and taking the time to coordinate can fall by the wayside. This is very difficult to overcome. New projects have to be implemented outside of the standard corporate culture in order to have any chance of success.

- **Project manager style**   Management may have encouraged an "explorer" image and style for project managers. They are treated as explorer Sir Francis Drake was—they are told to go out into the world with their project, to sink or swim. Using a collaborative approach on multiple projects undermines the autonomy of the individual project manager.

- **Nature of the work and geography**   The projects in the organization may truly be very different from each other. Moreover, they may be performed in different parts of the globe. In that case, you have to ask what they share, independent of tactical work and geography.

- **Lack of controls over project management**   While many companies endorse tools and methods and require some basic reporting structure for projects, there is often a general lack of controls. Few standards are enforced. Instead, people customize the methods and tools to a specific project and what they remember from the past.

THE PROJECT DIRECTOR

Managing multiple projects calls for skills and capabilities beyond those of a manager of a specific project. Instead of driving toward one set of objectives, the manager of multiple projects (here called the project director) must balance and trade off resources to prevent conflicts among the projects. The director must also allocate time and work at a different level of detail from a standard project manager.

How does one become a good project director? One way is to manage several projects at one time informally. This allows the manager to get a feel for how to allocate time, what the proper level of detail is, and how to manage and track issues, actions, and the project plans across multiple projects.

MULTIPLE PROJECT MANAGEMENT GUIDELINES

Several alternatives for managing multiple projects will be discussed and then we will focus on the one that has the greatest promise.

- **Alternative 1: Treat each project separately.**

  In this case, you receive little economies of scale. There is little chance of getting lessons learned since information is not shared. When faced with a resource conflict, management may be swayed by the best marketing job as opposed to the real situation. In addition, people may not even agree on the real situation, thereby confusing management more.

- **Alternative 2: Group all projects in one set and manage all of the projects together.**

  This is the micromanagement of all projects in a similar way. However, projects can differ greatly—even within a homogeneous company. Treating small projects like huge projects can swamp the smaller projects in administrative trivia.

- **Alternative 3: Establish several task forces (task group committees) to address projects at the group level.**

  Each task force is based on a grouping. Also set up an overall steering committee to address strategic issues and opportunities for all projects.

  The most important task group committee is the tactical resource group. This group deals with resources conflicts in the near term (1 to 2 months) and with dependencies between projects. A second group is the project management process group. This group addresses the development of project templates, methods and tools, and technology. These committees meet on a regular basis and produce specific actions and results. The groups of managers and staff act to implement collaborative decision-making. The steering committee addresses overall strategy and priorities among projects. It endorses the use of specific methods and tools.

- **Alternative 4: Use collaborative scheduling.**

  In collaborative scheduling, the project team shares, updates, and accesses the same project information. Project managers can even work with a common schedule. For multiple projects, this is opened up to leaders of all of the projects. That is, project leader A can read project B's plan and schedule.

Additional alternatives would be, for example, employing just the steering committee. This is a poor choice because too many issues and questions are referred to upper management. The task forces or task group committees act to buffer upper management, provide consistency, and allow for more detailed issues.

COLLABORATIVE SCHEDULING AND DECISION-MAKING

Across several projects, collaborative scheduling allows each project manager to do "What if . . .?"" analysis to determine trade-offs. With suitable project management software, a manager could, for example, extract five different schedules and combine them with a copy of his or her own schedule. Dates, durations, and resource allocation can be changed to see how to resolve a conflict. This is just the tip of the iceberg, since project leaders and key staff could view each other's action items, issues, and other project information, but each could change only their own data.

Once people are sharing the same information, they tend to arrive at a common view of reality with respect to the project. No longer is everyone coveting his or her own schedule. No longer are project managers having meetings in which people argue about status using different versions of the schedule.

With such common ground, it is possible to establish decision-making. Rather than have all decisions flow up to the highest level of management, collaborative decision-making empowers the project managers and their key staff members to arrive at solutions to resource issues, conflicts, and other problems. These meetings can be coordinated by the project schedulers.

The benefits of this approach are substantial. First, the elapsed time to resolve conflicts is reduced. Second, because the decision was reached by group process with the same data, chances for reversal later are small. Third, through participation, managers feel that they have a stake in the action and are more willing to work toward a solution.

EIGHT STEPS TO MANAGING MULTIPLE PROJECTS

The following steps can be taken to put into place a collaborative process for managing multiple projects:

Step 1: Determine project similarities and differences. In this step you attempt to understand all of your projects.

Step 2: Define approaches for grouping the projects. Employ the groupings suggested earlier in this chapter.

Step 3: Group the projects. This is the management decision as to what projects will be considered part of the same group.

Step 4: Define minimal standards for projects. This is the definition of what constitutes a set of rules for all projects. It can be graded in levels to projects of various size, complexity, and risk.

Step 5: Develop an analysis and reporting process across projects. Reporting on projects was explored in Chapter 13.

Step 6: Define a process for sharing lessons learned across projects.

Step 7: Resolve resource conflicts among projects. Decide to what extent you will implement collaborative decision-making.

Step 8: Implement a project management steering committee and the group committees.

Step 1: Determine Project Similarities and Differences

To get started, gather information on recent projects, current projects, and future projects. Classify the projects in terms of a series of attributes. Here is a table to help you:

Project	Importance	Size	Organiz.	Duration	No. of Key Resources

To fill this in, use subjective estimates. Importance in terms of impact on the organization might be low, medium, or high. Duration can be treated in terms of a time frame (e.g., small—less than three months; medium—up to a year; large—more than a year). Organization is the main organization benefiting from the project (not the organization performing the project). Size refers to project size in terms of budget (set your own intervals here). Number of key resources refers to the number of distinct types of personnel, equipment, and facilities important to the project.

You could add many more attributes in this table, but avoid getting buried in detail.

This table reveals some ways in which to arrange and sort the projects. You can determine the relative number of projects that fall into each category. This table and others have also proven useful in indicating the mix of projects to management.

Step 2: Define Approaches for Grouping the Projects

The table here is intended to identify the critical projects from the analysis of Step 1. In both the row and the column headings, put the name or identifier of the project. In the table elements, you can place several codes as follows:

- R— projects share resources
- D—projects are interdependent
- M— projects are under the same general manager
- O—projects affect the same organization
- P— projects address the same business process
- T—projects share the same technology
- MT—projects share the same methods and tools
- C— projects address same customers
- SU—projects address same suppliers

You are likely to have multiple entries in each cell. Your next step is to shuffle the rows and columns so that projects that have much in common are in the same block of the table. The result will appear as shown in the box below where the heavy lines split the four projects into two groups. Begin with grouping by resources as the primary criterion. Within this, group by business process. Note that the diagonal is blank since the row and column are the same and the table is symmetric. In this example, projects 1 and 2 share resources, business processes, and management. Projects 3 and 4 share a dependency and the same methods and tools.

1	2	3	4
R,P,M	O	T	
R,P,M		C	S
O	C		D,MT
T	S	D, MT	

Step 3: Group the Projects

Place the groups based on the analysis of the previous two steps. This will determine how you will manage the projects overall and how issues will be evaluated.

Step 4: Define Minimal Standards for Projects

Recall that you have a large range of projects, small to large. What are the minimal standards that you want all projects of any size to adhere to? If you impose too much, you will drive small projects underground. To provide flexibility, minimize the number and extent of standards that will apply across all projects. Minimal standards are required for the following areas:

- Project reporting
- Project and milestone review
- Methods and tools to be employed in each project
- Contents and structure of project plans

Step 5: Develop an Analysis and Reporting Process Across Projects

Each month do a rollup of all projects. To make this usable, insist on a standardized format. Let's first assume that the project plans for all projects are electronically stored and available on the network for access. Use the standard project report discussed in Chapter 7. Define the overall analysis approach to be employed across projects.

Here are some suggestions:

- Summarize the top milestones achieved in the last period.
- Indicate the top ten issues and their status and what is expected.
- Present a combined project plan that summarizes all of the individual project plans.

The combined project plan can be obtained by rolling up detailed tasks into summary tasks and summarizing these, and by combining the summarized projects into a single project.

Step 6: Define a Process for Sharing Lessons Learned Across Projects

Based on groupings in steps 1 through 3, you can determine which projects can share specific lessons learned. Each experience or lesson learned can be classified as to technology, method, tool, organization, etc. Employ this information to identify the projects to which the lesson learned is applicable. If you employ groupware or a database as a means of storing and accessing lessons learned, then managers and staff can access the relevant items for specific issues or topics. In follow-up, evaluate whether the lesson learned can be expanded after it has been applied and make the effort to see if the benefits were achieved.

Step 7: Resolve Resource Conflicts Among Projects

Using collaborative scheduling and management, the approach would be to gather project leaders in a room for a group of projects that will have resource conflicts over the next four weeks. Have a moderator, such as the project scheduler, first introduce each issue and conflict. If possible, project the computer screen on the wall of the room so that everyone has a visual of the conflict. If you combine the detailed schedules of the projects in conflict, you can perform "What if. . .?" analysis by assigning resources to some projects while allowing others to slip due to lack of resources. By trying out several alternatives, the project managers will be able to arrive at an agreement on what to do. The schedules can then be changed to reflect the results of the meeting.

What are some alternatives to this approach? A traditional method is to have project leaders go to their own managers to seek priority for the resources. This entails more meetings and coordination. Management at higher levels must be involved. This method was used at the consumer products firm before collaborative scheduling. The drawback was that this method consumed too much management time. Decisions often were based on the marketing skills of the project leader.

Step 8: Implement a Project Management Steering Committee and the Group Committees

The steering committee is a management committee that oversees all projects. It reviews and approves new projects. It can terminate, merge, and redirect projects. The members of the steering committee represent a cross-section of the organization. In some organizations this is called a technology committee; in others it is called a process committee. The committee provides a company-wide perspective on all projects.

Disputes and issues that cannot be resolved within the project or within the conflict-resolving process are presented to management through the steering committee. The committee also reviews all new project ideas.

Benefits of the steering committee approach include the following:

- Upper management does not have to be involved continuously in project issues since the committee represents a forum for them.

- Projects tend to get treated more fairly, since most of the units of the company are represented on the committee.

- The committee represents an appeals process if the deconflicting process does not work.

- Upper management tends to be more interested in and excited about the projects with the committee in place.
- The committee provides a structured approach for dealing with new technology and reengineering.

Both the construction and consumer products firms implemented the steering committee approach. In the former, it was expanded to include procurement of major components and systems as well as the construction projects. In the latter, it was expanded to include all new technology in support of product development.

What do you have when these steps have been completed? Here is a summary list:

- A standard set of project templates that can be used as the basis for most new projects
- Minimal standards that apply to all projects, along with guidelines to be applied to projects based on risk, size, complexity, and other factors
- A process for reporting to management on all projects
- A summary reporting and analysis method
- A process for gathering, storing, accessing, and disseminating lessons learned
- A process for resolving conflicts between projects (deconflicting)
- A project steering committee for overseeing projects

MANAGING RISK ACROSS MULTIPLE PROJECTS

If the projects are interdependent, you can combine them into one large schedule and attempt to find the management critical paths. Risk and exposure are to be found in areas such as technology, the process, methods, tools, organization, and management. In other words, risk tends to reside in the groupings that were developed earlier. Therefore, to assess risk, identify the group of projects affected by the area of risk. Assess risk as a group by considering slippage and problems within each project.

EXAMPLES

The Railroad Example

In the case of three of the railroads—the Union Pacific; the Atcheson, Topeka, and Santa Fe; and the Central Pacific—the need for parallel projects was critical. Otherwise, deadlines imposed by financial considerations (running out of money), the weather (winter), and government aid (subsidies) might run out. Each of these railroads and others had multiple surveying, construction, support, and logistics operations going on in parallel. These projects were all coordinated by a head engineer.

Allocating people across the multiple projects was a challenge due to the difficulty in communications and support. For this reason, a core team would be established in a particular area. As work progressed, more men would be sent to the site. This would trigger the emergence of a temporary railroad town. Records show that the railroads began to move people between projects as the railroad line expanded, especially when repairs were necessary or to get full benefit from people with knowledge and experience. Portable sleeping cars were used to support this mobility.

In the case of some eastern and European railroads, the push for the railroads came from companies involved in coal, steel, or other manufacturing. Railroads offered a less expensive and quicker way to transport goods than rivers or overland shipping. Prior to railroads, the mine owners were held captive by the river shipping companies. The railroads turned them loose. Companies in Europe and the United States involved in natural resources were able to expand into support of both the mines and the railroads.

Coordination among railroads did not occur directly in the United States until the latter part of the twentieth century. Customers complained about filling out separate shipping forms and documents for each railroad. If their goods were to traverse four different railroads, they had to complete four sets of forms. In response, the railroads worked together with the government to standardize the forms. This standardization carried over to computers and the electronic interchange of information and was one of the earliest examples of Electronic Data Interchange (EDI).

Modern Examples

The project manager at the manufacturing firm started out with a single large project structure. It became clear that this was not working efficiently. Some countries were moving ahead much faster than others in network installation. If this structure had continued, more aggressive offices would have been held back. The projects were eventually broken up by country. Each country had a project manager. All of these projects shared a common template, methods and tools, and technology. The schedules were consolidated into an overall summary.

Construction projects can be grouped in several ways: geographically, by type of construction project, by customer, and by shared resources. After struggling with this for some time, the construction firm decided that the best approach was to maintain these different groupings because the projects were quite fluid and change was rapid when it occurred. Templates were constructed for each type of project. At first, resource conflict was addressed by project managers using electronic mail and telephone at first. Later, experiments with limited videoconferencing proved promising. However, the network speed made the videoconferencing clumsy.

GUIDELINES

- Put external comparisons with other projects into place early.

  Begin to benchmark projects early so that you can track where you are going and how you are doing. This also helps motivate the project team as they see how they are doing vis-a-vis the other projects.

- At the start of a project, determine what controls, methods, and tools are appropriate to the project.

  Treating all projects the same will smother small projects and let large projects creep out of control. Projects and project work are not democratic institutions.

- Take control of the resources only when you are doing actual work.

  Some project leaders use project schedules to hoard resources. This problem will surface immediately when you address resource conflicts.

- Use different teams on successive projects.

  A common argument goes that if a team can succeed in a project, the team should be moved to the next project. But the same team may not be successful on a different project. Also, the team may have developed ties that are too close and that can disrupt the work of new project members. Also, keeping an existing team intact means that the next project manager will inherit baggage in terms of problems, mistrust, hatreds, etc.

- Identify projects in groups to support lessons learned.

A project that is based on technology is often plagued by gaps created by the numerous differing technologies, methods, and tools. This can happen with several projects that are going on in parallel. Because of limited communications between projects, it can happen that each project solves the gap problem differently. This duplication of effort can be eliminated by the sharing of lessons learned.

STATUS CHECK

- How does your organization address multiple projects that are not related? That are related?
- How does your organization roll up the overall projects?
- How does the organization cope with assigning and prioritizing resources between projects?

ACTION ITEMS

1. Develop the table in Step 1 to determine the resources and other factors required by a set of projects.

2. Next, develop the table in Step 2 that supports grouping. Proceed to Step 3 and group the projects.

3. Now evaluate how the organization is managing the group of projects. Are you taking full advantage of the common elements between the projects?

CHAPTER 17

ACHIEVE SUCCESS WITH ELECTRONIC COMMERCE PROJECTS

CONTENTS

17

ACHIEVE SUCCESS WITH
ELECTRONIC COMMERCE PROJECTS

INTRODUCTION

Many future IT and business projects will involve e-commerce. Electronic commerce consists of establishing electronic business relationships with customers and suppliers. In e-commerce the goal is to automate as much of the business processes as possible.

The purpose of this chapter is to explore how to integrate the methods and techniques in this book with the complexity of e-commerce.

Here are a few of the reasons that e-commerce is complex:

- To implement e-commerce you must involve many different parts of the organization. These have their own interests and agenda and they must integrate their processes, perhaps for the first time.
- There is often tremendous management pressure to implement e-commerce quickly.
- Many organizations must involve customers or suppliers in the implementation of e-commerce.

This complexity often leads to failure. However, e-commerce must be implemented to give companies competitive advantage and sometimes to ensure survival.

This is a challenge and failure to do this is also a major reason for failure in e-commerce.

There are many issues that you may face. Here are some examples:

- Internal managers may resist e-commerce because of a lack of understanding of e-commerce. This means that early in the project you must educate managers by example as to roles and challenges. Point out the downside of failure and of not implementing e-commerce.

- Managers do not think beyond their narrow departments. They refuse to sacrifice the traditional self-interest of their departments for the larger good of the e-commerce implementation.

- There is a tendency of keeping the current systems and to build new functions around these systems. This often does not work because the current system is not geared to online, interactive use along with interfaces with other systems. Modularity in systems is critical.

- It is possible to rely too heavily on consultants and contract help. After all, these people have experience in e-commerce that you lack. They are experts. However, you must implement e-commerce within your culture.

- People want to run out and buy some e-commerce software, thinking that they can implement this and then interface it to their existing software. This can be deadly unless the overall architecture for the entire e-commerce system is defined first.

- Companies are overwhelmed by the challenges. Structural problems in the organization that have existed for many years may have to be solved.

- E-commerce threatens well established empires in a company. Transactions flow across departments. The standard barriers between departments are blurred.

- The Information Technology (IT) organization may be resistant to changes and the pace of change. The IT group cannot be circumvented. It must be part of the solution. Simply hiring consultants is not a satisfactory solution. The members of the IT group have knowledge that is essential for success. The group needs to be augmented in terms of budget and staffing.

PURPOSE AND SCOPE

The purpose is to implement e-commerce processes and organizational structures that are responsive and supportive of e-commerce. Start with this wide view of the project, rather than looking at the goal as merely to implement an e-commerce system.

Next, consider different perspectives. For each you can define more specific objectives. Here are examples of objectives:

- The *technology objective* is to implement a technical solution for e-commerce that is flexible and scalable. You must have flexibility to accommodate the non e-commerce part of the business. In e-commerce, the workload increases with popularity, and thus you must make an effort to scale the use of e-commerce to the project itself.
- The *organization objective* is to support e-commerce transactions for the long haul—not just initial implementation. You must consider how your organization will work with suppliers and customers in an e-commerce environment.
- The *business objective* is to establish business processes that are scalable, controlled, and efficient. This requires more than just some reengineering work internally. The business processes must reach out to suppliers, intermediate firms, and customers.
- The narrow *management objective* is to position the company for the future. The terms *competitive advantage* and *profitability* apply here. When you implement e-commerce, you are setting the future direction and position of your company. E-commerce is more than just an extension of what you currently do.

In terms of scope, there is almost no limit. The scope encompasses organization, technology, business processes, procedures, and staffing. With suppliers and customers involved, the scope grows even more. A major challenge is to keep the scope of the project manageable.

Start with the internal processes, systems and technology, and organization. Soon you will add the suppliers and customers. Then consider competition, not only in your industry segment, but also from firms who have e-commerce already. They have been known to invade the turf of established companies. The boundaries between industry segments have blurred.

PROJECT PLANNING FOR E-COMMERCE PROJECTS

If you take the larger purpose, scope, and roles discussed above, you may think that implementation of e-commerce will take years. That was true some 30 years ago, but not today. There is too much pressure for rapid deployment and implementation. What should you do? Experience from successful projects points to a two-pronged solution. On the one hand you need a long-term strategy and plan. You also must have short-term implementation projects that support not only the long-term plan but also specific e-commerce transactions. Divide the work into multiple small projects. Each area should be defined as a separate project. Then you can create a master schedule of the individual projects.

Here are the separate project areas:

1. Technology This project deals with the hardware, system software, network, database management, and software utilities that will support e-commerce. The purpose of this project is to establish the infrastructure for e-commerce. The scope must include technology relationships with suppliers and customers. Issues you may face include the following:

- **New technology** You may need new technology. Your network may need to be upgraded to accommodate growth.

- **Interface to the current technology** Most people cannot afford to replace everything. So you are faced with interfacing to the current technology that will remain.

If the staff is not be familiar with the new technology, there will be a learning curve. You may have problems if you select the new technology without planning for the software and other parts.

2. Software Here you are concerned with the new e-commerce software. Identify all of the software components of your e-commerce solution. If, for example, you are setting up a retail web site, you would address user interface software, cataloguing software, and merchant software for credit card processing,

A common mistake is that people miss some of the necessary software. A second issue is that great attention must be paid to the integration of the software. This does not mean that you have to buy all of the software from one source. You just have to pay attention to integration. A third issue is the problem of scalability. Your software needs to scale up to the greater volume of work.

3. Interfaces   There are many interfaces to both existing and new systems. Internally, the e-commerce software has to interface to your order entry, inventory, shipping, and accounting systems—as a start. Externally, you may have to interface to supplier systems, to third-party networks, and to a credit card authorization and clearing firm.

A major problem and risk in interfaces is lack of knowledge of whether or not the interface will work. This is especially true or external interfaces. People like to work with what they know and are familiar with, and people tend to focus on the internal interfaces. The problem is that it will take a great deal of effort to define and set up the external interfaces. This area should not be ignored.

4. Business processes   Take a total view of your business processes. These include ordering, order processing, inventory, shipping, accounting, and others. You may have to establish a credit card processing operation. This must include fraud, reconciliation, and other functions such as refunds. You also may have to create a new process for customer relations.

The new business processes for e-commerce should be defined first. The risk is that people will attempt to overlay the e-commerce transactions over the existing transactions. When defining the new work, focus on each transaction. Don't just divide up the transaction by department.

Many people stop when they have defined the e-commerce workflow, but they should go on to integrate the e-commerce and traditional processes. You should simulate the workflow of both types of transactions to see how they are consistent.

5. Supplier and customer relations   This concerns establishing relationships with suppliers and customers. Define a strategy for dealing with both. This may be new ground for you and your staff. You may lack expertise on how this should be set up.

Often, people hire a consultant and turn over responsibility for this project area. However, you will have to operate and expand this area later. It is better that you and your staff gain the knowledge needed to deal with this, even at an early stage.

6. Organization   The departments are going to be shaken up as a result of e-commerce. You risk failure if you treat this as an add-on to the current organization. Analyze all of the line departments that are touched by e-commerce. You may have to create new departments.

There will be some resistance to a new organization. Many people may think of it as just another channel for business. This is one of the issues that you will need to keep in mind.

Pay attention to the interfaces among departments. Normally, you would concentrate on individual departments and their charters. With e-commerce, you might want to define the interfaces among the organizations first. Then you can move into each department. In creating a new or modified organization, you will be faced with integrating the e-commerce activities and the activities that do not involve e-commerce in each department.

Now move up to the overall project. Create templates for the areas defined above. You may want to consider creating another project that deals with interfaces among the projects, such as the following:

- **Business processes and organization**

  The business processes must be started first. As this project progresses, the organization project can begin. As both projects go on, a continuing sharing of information and integration will be needed.

- **Technology and software**

  Obviously, the technology must be in place to support the software. However, common planning for both areas is also required.

- **Software and business processes**

  While e-commerce software is sufficiently general, there must be a good fit between the processes and software. Otherwise, you suddenly will find yourself creating manual processes and work.

- **Organization and customer-supplier relations**

  Customer-supplier relations need to be addressed in the organization. It is logical to begin the customer-supplier relations project before the organization project. You cannot create a separate organization for supplier-customer relations. You need to consider customer or supplier aspects within many different projects.

- **Customer-supplier relations and business processes**

  These projects can be started in parallel. The two projects must interface. In particular, pay attention to how customer-supplier relations are involved in the business processes.

Now move to the overall project. Establish an integrated set of issues across all projects. One reason for this is that the issues are likely to be interrelated; many have ramifications in other projects. An even more important reason for this integration is that you will be able to use the common issues to encourage people to work together and share information.

METHODS AND TOOLS IN AN E-COMMERCE PROJECT

Individual-based methods and tools are not workable in an e-commerce project. Focus instead on methods and tools that support collaborative work. Consider database, groupware, and project management tools first. Keep in mind that replacing tools or methods during the project will be very difficult, given the number of people involved. Remember also that these tools and methods will be used after the project is finished. An e-commerce project is always a work in progress. It is really a program that progresses through stages as the e-commerce effort moves ahead. Select methods and tools that are based on common sense and can be learned and used by many different people over time, so that training is not difficult and time-consuming.

COLLABORATIVE WORK

An e-commerce project cries out for a collaborative approach. People must work together across the entire project. Information sharing and collaboration will increase if they have a good start.

To support collaboration, provide opportunities for it. This must be done outside of formal meetings. Hold informal meetings in which you focus on sharing lessons learned and knowledge and identify experiences that can be used in the new process. You can also identify issues related to organizations and processes.

MANAGING THE WORK IN AN E-COMMERCE PROJECT

A key phrase is *continuous coordination and cooperation.* The project team will spend much of its time resolving issues across the projects. Project reporting is also a critical factor. Management must be kept informed beyond status so that there is a continuous awareness of issues.

Another part of managing a project team is measurement. Measure the project and impacts as you go. One of the first tasks in each project is to undertake measurement of the current situation. As the project progresses, continue to measure not only for progress but also for impact.

Going beyond internal management, also assess external firms and what competitors and potential competitors are doing. A sudden change or emergence of a competitor can change priorities within the project. The way to deal with this and to avoid unpleasant surprises is to stay on top of the industry.

One major firm was late in getting into e-commerce. It finally got started and seemed to put on blinkers. When the site was up and running, competitors already had the same or better capabilities. The firm has not been able to gain ground.

Two other firms experienced crisis by acting too late or by treating e-commerce lightly. For one firm the result was the exit of the CEO. For a computer manufacturer, the only way to survive was to acquire another firm. The manufacturer overpaid for the new firm and has not integrated the new firm into the culture and processes of the existing firm.

EXAMPLES

A leading firm that manufactures and distributes machine shop equipment had had comfortable growth and success. However, it was missing out on an opportunity. More than 100,000 small machine shop operations with 50 employees or less were untouched by the firm. The company developed a strategy for implementing e-commerce with these new customers. The company managers realized that a passive web site would not be sufficient. They had to offer full service.

In terms of organization, the sales and support forces had to be changed. A group had to be formed to visit and establish relations with these new customers. The sales force had to be incentivized to go out and promote the web site.

The business processes had to be changed to accommodate the small orders along with major orders from established customers. They could not operate in parallel. The existing processes had to be integrated with e-commerce processes. In terms of technology, some systems could be interfaced to e-commerce software, while other software had to be replaced.

An overall project team was composed of high-level managers. Additional project teams were formed from employees. The project management process aimed at a full collaborative effort. Employees were encouraged to supply lessons learned and experience in setting up new equipment, operating the equipment, and helping machine shops bid on jobs.

The result was a major success. The web site today goes far beyond standard product catalogues. Users can review lessons learned. A bidding wizard supports proposal generation by machine shops. Morale in the company has risen. Information is shared to a much larger degree.

This example shows the advantages of being creative in your approach to e-commerce and its implementation.

Another example was a firm that already had a web site and had a retail catalog operation. The challenge was to turn the firm into an e-commerce company. The company acquired a firm to help create the software. All departments were reengineered to support e-commerce. A separate but related project was undertaken to change the culture of the organization. Marketing and sales offered products on the web site. This was a challenge because it required reaching out to entirely new suppliers. The number of products on the site has risen rapidly and online sales have grown steadily.

Success was due to management's total commitment to e-commerce, even to the point of risking revenue loss to get a good start.

GUIDELINES

- **Management commitment to e-commerce cannot be subtle or just an initial push.**

There must be management involvement in issues related to organization, business processes, and other areas.

- **There is a role for an overall e-commerce program manager.**

  Note that the word *program* and not *project* is used. This is because the work will not end with one project or a set of projects. As soon as some e-commerce transactions are set up, there will be a need to expand the range of transactions. There will also be a need to tune the organization, processes, and technology.

- **Concentrate on some intermediate results.**

  Before e-commerce even rolls out, there will be organization and technology changes. This is beneficial because the organization morale and efficiency will be improved. There will also a rising tide of collaborative work.

STATUS CHECK

- Are you e-commerce ready? Assess organization, business processes, staff and management, technology, and systems.

- What are some viable new strategies for e-commerce for your organization? Consider both aggressive and conservative strategies.

ACTION ITEMS

1. Undertake an analysis of your current processes first. The other areas are important, but the key to success is to focus on the processes.

2. Get a reading of whether managers are aware of e-commerce. Consider hiring someone to train the managers and open their eyes to the potential and challenges of e-commerce.

CD-ROM ITEMS

17-01	Project Template for E-Commerce
17-02	E-Commerce Project Checklist

PART IV FOUR

PROJECT CRISIS

CHAPTER 18

DEALING WITH CRISIS

CONTENTS

18

DEALING WITH CRISIS

INTRODUCTION

The word "crisis" may sound ominous. However, a crisis can be viewed as an opportunity. It is a chance to gain management attention, to get decisions made and implemented, and to redirect the project. In reengineering projects, a crisis is employed as a trigger to change the organization. In other words, use the crisis to address a number of political, organizational, policy/procedural, and technical issues at one time.

When does an issue become a crisis? This is subject to interpretation. Some of the major factors affecting whether an issue has become critical are the following:

- Current state of the issue
- Rate of decay of the situation
- Increase in impact in the project or organization
- Age of the issue—how long it has been active

Crisis is a matter of perspective. Issues typically become critical to the project team first, then they escalate and progress through the organization.

What can you do about a crisis? You can understand and solve it. You can also guide and orchestrate it. You can play a role in defining the timing and presentation of a crisis, affecting the media surrounding the issue. That is, a crisis is open to interpretation.

The basic strategy in dealing with a crisis is as follows:

- You can affect and impact how a crisis is handled and resolved. Therefore, be proactive.
- A crisis can be employed as a tool to carry out fundamental change.

PURPOSE AND SCOPE

The purpose of this chapter is to provide techniques to address crisis situations so that not only is the crisis resolved, but also the crisis is resolved to the advantage of the team, the organization, and the project.

More specifically, the goals are as follows:

- To help you sort through the perception of a crisis vs. a real crisis
- To provide tips on how to analyze and assess a situation
- To support you in using a crisis to the advantage of the project

The scope of this chapter includes situations such as technical problems, organizational upheaval, project leader change, and major team changes, as well as more classical budget and schedule crises.

APPROACH

A crisis is a culmination of events that forces the project management, and organization to deal with issues. Thus, when issues become critical, you have a crisis.

A crisis in a project is a situation that requires rapid decisions and actions. If action is not taken and the crisis is real, the situation worsens, impacting the cost, the schedule, the quality, or some key attribute of the project.

A significant point to keep in mind is the difference between perceived and actual crisis. In a project it is important to act upon perceived crises as well as actual crises. Otherwise, if people perceive a major problem and you do nothing as a project manager, you and the project lose. You can use the perception of a crisis to achieve a breakthrough in the project in terms of resources or other factors.

A crisis typically begins as a cloud on the horizon. Suddenly more clouds appear and people sense trouble. Dictionaries define a crisis as "a turning point in the progress of a situation" and "a state of affairs in which a decisive change for better or worse is imminent." A crisis is subject to the interpretation of the audience.

If you announce that some situation or issue is a crisis, you have to back up your statement. If, on the other hand, you begin to point out that the impact of an issue is getting bigger, and risk and danger are growing, people get the impression of urgency without the word crisis. That is the preferred strategy. Save the term *crisis* for true emergencies.

When people see a project crisis, it is usually because they perceive an impact on their organization, the project, the project team, or some project resource. This perception can stem from internal as well as external factors. Thus, you must consider how these perceptions arise, how to determine if the crisis is real, what to do about it, and how to implement decisions.

Over the years, it has been found that certain projects are prone to crises. Here are some examples:

- The project is large or involves many people. This makes the project more visible and subject to misinterpretations and rumors.

- The project is political. Reengineering projects are typical here. Enemies of the project create crises based on some event.

- The project involves outside entities or external factors that can affect the project. An example is a project to remodel retail stores. The success of the project is subject to trends in industry and technology, as well as to what competitors are doing.

- The project extends over long periods of time and is exposed to many different factors.

CRISIS ASSESSMENT

Symptoms of Crisis

Symptoms are visible signs of the crisis. They may or may not be related to the actual causes. When you are sick and go to a doctor, the doctor is trained to observe the symptoms as a means of diagnosing the cause. Treatment is applied once the cause has been found. In the last few decades, doctors have been taught to listen more carefully to patients and to spend more time analyzing the symptoms, as opposed to rushing to diagnosis and treatment. This technique also applies to projects. Spend a generous amount of time assessing the symptoms. If the manager is remote from the project, symptoms typically surface as part of a review.

Here are some common symptoms of crisis:

- A lack of decision-making or only partial decisions are made
- Attempts to leave the project team
- Overrunning the project budget
- A lack of enthusiasm for the project
- Unresolved important issues
- Excessive calm as people try to ignore the crisis
- Excessive excitement as people address the crisis

These and other symptoms are usually visible if you are on the project team. However, the team members may be so accustomed to the symptoms that they do not even notice them after awhile, especially if the symptoms are not visible outside of the project. Other major sources that may alert you to a crisis are management and someone outside the project.

Predicting a Crisis

Many times you can predict a crisis. Although a crisis can appear suddenly it is more likely due to the impact of an unresolved issue that grew. You may be aware of the situation and issue. You may have pointed it out to management but nothing was done. For example, the schedule may reach a critical point and the project begins to fall behind. However, critical resources are not approved for the project. Eventually, a schedule crisis results. Because you predicted the crisis, you would have been able to prepare for it early.

Here are some guidelines in making a valid prediction:

- Take a broad perspective over time. This allows you to see what has been accomplished in the project, the rate of progress, and trends in the project.
- Look at how long important issues have remained unresolved.
- Observe how the project manager and the team have dealt with previous situations.
- Think of other projects in the past that are similar to this project.

An exercise at the end of the chapter asks you to take a project and define potential crises. It is important for you to sit down at least once a month and assess potential crises in your project. Then attempt to figure out what countermeasures you would take.

Causes of Crisis

Once you understand the symptoms of the situation, you can begin to sort out causes. Usually some combination of factors contributes to most crises.

Here are categories of causes of a crisis:

- **Political**   Someone or a group is out to sabotage the project. The group may not even care about your project; people may just want your project's resources and money for their own project.

- **Technical**   An inherent technical problem or flaw in the systems and technology in the project should be addressed.

- **Managerial**   This could be management indecision or a management vacuum. Whatever the cause, the symptom is managerial in nature.

- **Organizational**   The organization structure and roles inhibit the employees from coping with the project issues.

Here are some more specific causes of crisis:

- **Policy**   Is the crisis due to a new or existing business policy? An example might be a crisis resulting from a new approach to accounting or budgeting. A crisis may also arise related to how people are allocated to projects.

- **Internal project structure**   The basic project is based on goals and scope that are now less relevant. The project did not change relative to the evolving goals and scope.

- **Resources**   The resources in the project are not performing as expected and cannot address the problems.

- **Management**   Management attitudes or positions toward this project or other projects have changed. There is less support.

- **Project leader** The project leader may be the problem. The crisis is occurring because the project leader is not doing the job. Perhaps the project leader has let issues slip or did not validate the quality of the work.

- **The work** The quality of the work is not adequate. The work may not be complete. There may be a requirement for extensive rework or additional work that causes the schedule to fall apart.

Use a process of elimination when working with the above causes, since more than one cause can exist. Start by assuming that something in all of these is contributing to the problem.

COPING WITH A CRISIS

The First Steps

Let's move from the general discussion to action. Suppose that a crisis or important issue is before you. What steps should you take to address it?

Step 1: Determine whether the crisis is real or perceived. Ask yourself, "What has changed to make this situation a crisis?" Another test is to ask, "If nothing is done, what will happen? How will things worsen?" If your answers to these questions are that nothing has changed and the situation will not likely deteriorate, you have a perceived crisis.

Step 2: If the crisis is real, determine the scope of the situation and how much time you have left to make a decision and to implement it.

If the crisis is perceived, not real, decide what is behind the perception. Why do people feel that a crisis exists? Don't take action until you can answer this. Also ask who benefits if there is a perceived crisis. The perceived crisis may be a result of a misunderstanding. The analysis should reveal some interesting communications paths of the project. Take this opportunity to learn more about informal communications in the project. Determine the action that is appropriate. If you attempt to deny that a crisis exists, you will not be credible. Instead, think about what you want to say and the scope of your response. This is especially pertinent if you wish to use this opportunity to advance the project politically or to gain resources.

For example, go to management and indicate that the situation is bad but not hopeless. Identify what should be done before a crisis does erupt. This approach has several advantages. First, you are not crying wolf. Second, you are warning management before the crisis arises that the possibility exists.

Step 3: Implement the decision.

Here are some questions to answer for a real crisis:

- What other issues should be addressed as part of the actions involving the crisis-related issues? Handle as many open items as you can. It is important to determine the scope.

- What are the long-term approach and solution? Knowing these will ensure that short-term actions do not negate the long-term solution.

- What are alternatives for short-term actions? How do these alternatives group together?

The Strategy of Doing Nothing

Doing nothing is a reasonable alternative to action. Most of the time you should wait and organize a response to the crisis. Taking action without thinking through the consequences can make the situation worse. Time allows a situation to mature. Time allows you to work behind the scenes to get the situation resolved. Time also increases anxiety so that people will welcome a resolution and decisions that they would have resisted several weeks before.

Let's examine the classic case of a software development project. The project slipped. Management overreacted by throwing more people into the project. The people already in the project suddenly had to stop working and bring the new members of the team up to speed. The project slipped further behind. The lesson learned here is that a more intelligent approach would have been to reduce the size of the team and increase accountability.

Here are some events that can occur if you leave the crisis alone:

- If a crisis is emerging, the existing symptoms may become more acute. Additional symptoms may appear. The problem will worsen.
- The issue may be noticed by someone else who may address it before it becomes critical.
- The situation may not change at all. This does not indicate that the issue is solved. It typically means that nothing is visibly new about the situation.
- The problem symptoms may abate or disappear with time.

Our basic recommendation is to let an issue mature and emerge on its own. This allows you time to plan on what to do. This strategy also will help convince people of the urgency later.

How do you protect the project if you decide not to do anything? Alert management to the issue and indicate that you are watching the situation and planning for action. Point out that taking action may not be needed and you don't want to overkill the issue.

Inaction may appear to be a weak way to handle a possible crisis. However, you do accomplish something, even by waiting to act. First, people are on the alert. Second, you will be able to judge the scope of the issue more precisely. Third, you will have more time to rally management behind a course of action.

Traditional Management

In traditional project management, the project data and schedule are often tightly held by a scheduler and the project leader. Few people are aware of what is going on, since the projects are managed separately and are only loosely coordinated. This can set the stage for disaster, as a project can slip or an issue can remain hanging for weeks. When management is finally aware of the problem, too much time has passed and the action that must be taken is too drastic. The decision is rushed.

Without open sharing of project information, little chance exists of getting help from other projects. This also tends to make the situation worse. It is more difficult to approach other managers for ideas, given the closed nature of the process.

In employing the traditional approach, you are often forced into using the written word, be it electronic mail or memorandum, to indicate the problem. Going on record in writing can lead to trouble, as the problem is now out in the open.

Collaborative Management

In a shared information environment, all of the project team members, as well as other project leaders and members, have access to the issues and the project plan. This is an environment that encourages people to offer suggestions and help. People feel that they are working together even though they are on separate projects.

Here are several advantages of using the collaborative management mode in handling a crisis:

- The team members work together to head off the crisis, thereby building and reinforcing the strength of the team, as well as solving the problem.
- A crisis tends to be identified earlier and addressed earlier. Therefore, it is less likely to turn into a serious crisis.
- The factors in one project that are creating havoc are often present in other projects. With the shared information the issue can be addressed systematically and not ad hoc on each individual project.
- Team members who have worked on other projects may have experienced similar problems and can give good advice.

Dealing With a Crisis Using Modern Tools

The modern tools discussed in earlier chapters include collaborative project management in a network, electronic mail, groupware, and shared databases. Several basic features and effects of the new software contribute to its usefulness during a period of crisis, including the following:

- Information is shared and available to more people.
- People get the information faster than with other means.
- You can respond faster so that the volume of mail, messages, or transactions per unit time is increased.

With all of the benefits these tools bring, they also have a potential drawback. First, misinformation or incomplete information can be spread faster. Second, if this occurs, more people find out about it. It is difficult and in some cases almost impossible to correct false impressions. Third, team members must have access to the tools, be trained, and be active users of these tools so that they can receive the messages.

What is a good strategy for using technology to deal with a crisis? During the period leading up to a decision and action, minimize electronic mail and other media. Leave the project plan alone until decisions are made. After a decision has been made, take the initial actions without the electronic tools. Once these have a beneficial impact, start using the electronic tools again. Update the issues database and project plan. By then, the crisis will have passed. In summary, use the tools when you have more control of the situation and its outcome.

The Consequences of a Crisis

In a true crisis, management and the project work will be impacted while people deal with a crisis. Let's consider potential impact.

- **Paralysis of the project**   People are waiting until the situation is resolved.
- **Deferred decisions**   Decision-making is slowed.
- **Withdrawal of support**   People outside of the project who supported it suddenly are quiet and almost hidden.
- **Departure of some team members**   As a crisis continues, people start to jump ship.
- **Sudden increase in management involvement**   Due to the situation, management is now heavily involved in the project.

The positive impact a crisis has on a project is that it gives the manager a chance to redirect and reenergize the project.

IMPROVE YOUR CRISIS MANAGEMENT SKILLS

Hone your skills prior to an actual crisis. One of the hardest but most important skills to develop is the ability to have perspective and patience when confronted with a set of symptoms.

Here are some suggestions to help you accomplish this:

- Examine your list of outstanding issues. Ask if these are complete. Add to the list any politically sensitive issues.
- With the complete list, note the age and importance of the issues. Sort the issues in order of importance. Within a group of issues of the same importance, sort by age.
- Take the top five issues in importance and age and assess what the trend has been for each. Has the issue deteriorated, remained the same, or improved?
- For each issue, develop a scenario for the reasonable worst case. That is, how could each issue turn into a crisis? What would be the symptoms of a crisis? What would cause the issue to worsen to this extent?
- Now determine how you would detect deterioration in each issue.
- With an assumed crisis, attempt to develop countermeasures you could take.

Following these suggestions will help to prepare you for a crisis. Get in the habit of evaluating issues each month to raise your level of awareness of the possibility of a real crisis.

Step 1: Determine the Crisis Issues to Be Included in the Decisions

Include as many issues as possible. You want to fix as much as you can so that you can stabilize the project. Typically, one or two issues that must be addressed are obvious. What is not as obvious are additional political and organizational issues.

Here are two examples from projects.

- The project crisis is that the schedule and cost are getting rapidly out of control. The cause was not the work itself, but the fact that the scope of the project had been expanded slowly to include additional work. What decisions are possible? First, you can create a larger project plan and team. You can request additional funding. The drawback of these plans is that they may slow the project while they are being put in place.

A second course is to curtail the scope of the project immediately. This might help the core of the project, but it will create enemies of people who were expecting the additional items that were promised.

A third approach is to create a new super project and divide the project into subprojects. The core activities of the original project might be divided among several subprojects.

Which is the best approach? We are in favor of the third approach, even though it involves the greatest change. It requires changes in the plan, budgets, and staffing. You may encounter political resistance. You may be faced with management turning you down. Then you will have to revert to one of the other two alternatives.

- The project crisis is that an outside factor you cannot control is impacting the project. This could be another project or an outside event that is changing the nature of the work and purpose. The symptoms of this are typically doubt and tension within the project. What should you do in terms of defining issues? First, include the immediate impact on the project. Next, sit down and estimate what long-term impact there is on the organization and on the project. Attempt to widen the scope of issues so that they can be dealt with on a higher organization level than the project. If something is beyond your control, escalate it so that management will handle it.

These two common examples will be employed in each step. Note that within each are underlying political issues and also factors beyond your control.

Step 2: Define Possible Decisions and Their Interrelationships

In the first example of the budget and schedule overrun due to expanded scope, three possible options for handling the crisis were offered. With the third option, further decisions are required. First, decide how to divide up the project. Then decide if you are going to expand the scope even further to reach some natural boundary. This example shows that after you define a series of alternative major decisions, you next need to identify smaller decisions within each major decision.

When making lower level decisions, consider including at least one decision in each of the following areas:

- Project purpose and scope
- Project organization
- Interfaces outside of the project
- Project team and staffing
- Project methods and tools
- Approach to communications with management and outside of the project

Are some of the lower level decisions the same for several alternatives? If so, this may cause you to rethink the alternatives.

Here are examples of possible decisions.

Choose to do nothing. You will not make any decisions now. Instead, you will monitor the situation for changes.

Use no new resources. A second possible decision is to take no action that would require additional resources. You will live within your existing means.

Change purpose and scope. You will either expand or contract the purpose and scope of the project.

Work the political circuit. You won't change the project. Instead, you will go out and position the project and work politically. This method is useful when you have a crisis in interpretation of the project.

Add resources to the project. Determine if the project could benefit from additional resources to address the crisis. Consider this only as a last resort. Substitution of resources is included in this decision.

Modify the methods and tools in the project. This is also dangerous. If you change the methods and tools during the project, you will slow the project and cause the need for more time spent in training.

When considering subsidiary decisions, assess timing and dependencies. What would be the trigger for a later decision? If decisions are dependent, you may need to announce the decisions at the same time, but implement them sequentially. That is, you might change the project scope and the structure of the plan. Changes in resources would also follow.

Step 3: Determine Potential Actions

You have now identified the range of possible actions required to implement a decision. Take each decision and fit it into one or more of the categories in the following list:

- Changes In the Budget
- Changes In Staffing
- Changes In the Project Team
- Changes In Objectives and Scope
- Changes In Methods and Tools
- Changes In Organization
- Changes In Technology
- Changes In the Schedule and Deadlines

The action itself may involve a combination of activities, such as announcements, analysis, reorganization, procedure, policies, staffing, training, facilities, and equipment. It is important that you think an action through. If you neglect some parts of an action and people become aware of it, your credibility will be questioned. Once this occurs, the lack of confidence in you will expand to the entire action, as well as to the decision itself.

Step 4: Make and Present the Decisions

Three sequential events are involved in this step:

- Making the decision
- Announcing the decision
- Taking action based on the decision

To be most effective, the time between each of these events should be very short to prevent any buildup of resistance or short-circuiting by a manager. Rapid progression of events also reinforces the impression of you being action-oriented. If team members receive an announcement of a decision and this is followed by silence and inactivity, doubt and anxiety are created. Progress on tasks may slow.

Making the Decision

To prepare for making a decision, compare and analyze alternatives. Here is a table for analysis and for presentation to management.

Major Decision	Supporting Decision	Impact	Benefit	Risk	Comments

In the columns for impact, benefit, risk, and comments, you can enter information in a bulleted format. Use impact to enter what the expected result is. Benefit is the effect on the project and organization. Risk is the potential exposure and likelihood of problems.

Each decision or supporting decision has actions that must be taken to support the implementation of the decision. An additional factor that can be added for each supporting decision is the trigger or event that led to the decision being made and the action being taken.

Use a second table to record this information:

Major Decision	Supporting Decision	Actions	Trigger

If you want to pursue this further, create detail for the action items. This detail might include the following:

- How the action will be taken
- Who will take the action
- What anticipated fallout might occur due to the action
- Verification that the action was taken

In some cases, you will recommend a decision to management. In others, you will be empowered to make the decision yourself. In previous chapters on communications we discussed dealing with marketing and selling decisions to management, the team, and the organization. If you seek management approval, you have the above analysis in hand. The more thoughtful the presentation, the more support management will provide.

Who should make a decision? Many would say "Pass it up the ladder to management. They can make the decision and people will follow." If you do this, follow these guidelines:

- Base the level of management required for the decision on importance and scope to the organization. Have the decision made or endorsed at as low a level in the organization as possible.

- Inform managers at higher levels in advance of the decisions and actions. This has several benefits. The informed managers can impact the decisions or actions, if they feel that it is necessary. Also, the managers put distance between themselves and the decisions so that they feel less at risk. With advance notice, managers are better prepared for any issues that may arise. Give some examples of negative reaction that may occur due to resistance to change.

Announcing the Decision

Begin by making your announcement to the project team and work your way to the line organizations. In terms of timing, attempt to make the announcements as close together as possible.

With the announcement, suggest some actions to put into effect right afterwards. These might include project changes, team assignment changes, method adoption or change, or tool acquisition. You want to shift the attention of the people from the decision to the actions. This gets people working again.

An announcement should be preceded by a verbal announcement indicating what is coming. This reduces anxiety and stress and also preserves the surprise of part of the announcement. Never call off an expected announcement. This can raise more issues and doubts.

Taking Actions Based on the Decision

If a decision is not going the way that you think it should, don't continue to press for action. Try to put the decision on the shelf. What if immediate action is needed? Shift gears and work on selling the actions as a stopgap measure in order remove the pressure to make a decision.

Step 5: Implement the First Actions

The first actions are often the easiest to undertake. These don't involve purchasing or hiring and are probably within your span of control. You have considerable discretion as to how to implement these actions.

Here are some suggestions:

- Implement actions in groups, with periods of calm between groups of actions. This allows you to assess the effects of the actions.

- Determine what results should flow from the actions.

- Follow up on the actions to ensure that they are being carried out and that the results are what you anticipated.

- Be ready to step in and modify the action in terms of implementation, if necessary.

- Evaluate the actions individually and as a group.

When implementing an action, play an active role. If new procedures or announcements are necessary, review the material prior to release. Show that these actions mean something to you. If others get the impression that you don't really care and that this is not serious, they may not participate.

Be flexible with respect to actions. While changing a decision may result in actions changed and an unraveling situation, the consequences of changing an action are less far-reaching and may be advantageous.

Step 6: Measure the Results of the Actions and Determine What Further Actions Are Necessary

When you measure the results of an action, pose the following questions:

- Is the work on the project continuing?
- Has the nature of the work changed to reflect the decision and actions?
- What new factors have emerged that should be addressed?
- Do areas of ambiguity or fuzziness need to be addressed?
- Are the results of the actions supportive of the decisions?
- Is the pace of the results consistent with expectations?

People often forget management at this time. Once management gives approval, people go off and implement the decisions and actions. Seldom is there feedback to management on what happened. However, the project will progress more smoothly if you provide feedback to management after the actions have been taken. Keep management informed to retain support for the project.

Decisions, actions, and change can bring harmful side effects. Progress may slow. Other projects may be impacted. Sometimes the actions have not been thoroughly considered. Gaps exist, unanticipated problems arise, and issues are left unanswered. These situations can create enemies. Enemies can then go to management and indicate that all is not well with the project.

Keeping management abreast of what is happening will prevent management from relying on a project enemy for information, misinterpreting the situation, or feeling betrayed and turning against the project, all of which could result in management stopping your planned actions.

This discussion provides further support for taking actions immediately after the decisions. Measurement and marketing of the results must begin soon afterward. Adopt a conservative approach to measuring and marketing the results of an action. Figure out the worst case scenario and be prepared with solutions. Head off misinformation and disinformation on a positive note.

EXAMPLES OF DECISIONS

Replacing the Project Manager

Suppose that the decision has been made to replace the project manager. Short of terminating the project or a major redirection, this is one of most significant decisions you can make.

Once the decision has been made, the project manager should be informed. What happens next? We recommend a gap in management. Take time to consider what qualities you want in the project manager. An upper level manager can function as the acting manager of the project.

This approach has the following benefits:

- The interim manager can determine the true status of the project.
- The team gets a breather and an opportunity to tell an outsider what is really going on.
- Project changes can be made with the blame falling on the interim manager, thereby shielding the next official project manager from tough decisions.
- This gives a chance to test team members in assuming aspects of management of the project.

Recruiting from within a project may be easier than recruiting from outside. An insider already knows what is going on the project. No learning curve slows the project. The downside of recruiting an insider is that the person has existing relationships that are now changed. This creates new dynamics. If someone from outside the project is selected, the team members may resent the newcomer and feel that one of the current team members should have been considered. In light of the pros and cons, we recommend that you consider several team members for the position before turning to those outside the project.

If you decide that you must recruit a new manager outside the project, it is essential to find someone who will be able to take over an existing project and not change it to fit his or her preconceived ideas. Key qualifiers are that the new person is an "adapter" and is "results oriented." Also, choose someone who is a fast learner and doesn't require weeks to get up to speed on the project.

Terminating a Project

Beyond the analysis of the project itself, here are some factors to consider when making the decision whether to terminate a project.

- What is the impact on the organization of terminating the project? The project was to deliver benefits to the organization. These will not be received if the project is terminated. What is the impact of the loss of those benefits?

- What can substitute for the project? If the benefits from your project are needed, does another project exist that will result in the same benefits?

- Where can the resources best be redeployed?

- How should the project be terminated?

- What is the impact of keeping the project going for awhile longer?

Note that the following question was not asked: "What is the impact on management credibility or morale?" Do not keep a failing project going only to preserve management credibility.

To terminate a project, you can gradually reassign resources and pare down the project. However, this gradual death is often unproductive. If you need something in particular from the project, create a new task plan and have people moved to work on the necessary tasks. Terminate what is left.

Overcoming Resistance

You may encounter resistance to ending the project. Head this off by stressing the positive results that will occur with reassignment and work on other projects. Recognize the loss that people feel when they have worked diligently on a project that is scheduled for termination.

If team members still resist ending the project, try to convince them that this is the best course of action. Indicate the type of resources, changes, and other support that would be needed to continue. Also, stress what could be achieved in other areas by stopping this particular project now. If you are overruled and the project continues, don't say "I told you so" if the project eventually fails.

Conversely, if the project does turn around and is successful, trace this success back to changes made during the time of crisis.

How Should a Project Be Terminated?

Expediency is recommended. The faster resources can be moved, the less time the negative impact will be felt. Take the opportunity to build up the projects that will now receive additional resources.

Have a plan prepared to shut down a project before you announce your intentions.

Steps to Shutting Down a Project

Step 1: Freeze all documents and computer files. This will prevent people from destroying project information or end products because they are upset.

Step 2: Get people together from the project team and discuss lessons learned from the project. This allows people to focus on something positive.

Step 3: Let people know one-on-one where they will be going. Let them know what they will be doing.

People on the project team may blame themselves for the project failure. Even if they are to blame, tread lightly. Your goal is not to exact retribution but to achieve a smooth shutdown.

How long does it take to end a project? Your goal is to complete the process in a week or less. If lingering tasks have to be done, try to reorganize them so that they become part of a separate small project.

EXAMPLES

The Railroad Example

Crisis was a way of life for the railroads. The most common problem was lack of funds to continue construction. Since this was a recurring crisis, an entire financial industry, namely investment banking, was born. This expanded to handle other projects and industries requiring large amounts of capital.

A second problem was attracting settlers and getting firms to use the railroads—generating business in a wide open country. The most successful railroads expanded the scope of their firms to include sales, agriculture construction, and even steamships. Firms that recognized that a solution was necessary for a period of decades moved toward an expanded, integrated firm.

For the Central Pacific Railroad, a major problem was finding laborers to work on the railroad. People were not available due to gold mining and other industries. A crisis was at hand. The chosen solution was to import foreign laborers in very large numbers to work on the railroad.

The transcontinental railroad was merely a dream between the 1830s and 1840s. In the 1850s, some firms actually attempted to begin building it. However, these efforts were futile. The scope of the project was too broad. No individual company or firm had sufficient capital to make it work. The payoff to the railroad firms was too far off. Investment was not available because America was still an agrarian economy.

The American Civil War changed all of that. The importance of railroads to both the North and South was obvious. Abundant evidence showed that using the railroads had turned the tide of battle by delivering fresh troops and supplies to a battlefront. In fact, many of the Civil War battles were fought along railroad lines or around major railroad centers.

In the 1850s, a major battle had taken place in Congress concerning the route of a transcontinental railroad. Should it go via a northern route or southern route? No action was taken. During the Civil War, with members of Congress all hailing from the north, Lincoln saw a chance to get what he preferred. He proposed funding for the railroad and a northern route.

Lincoln saw several benefits to building a transcontinental railroad. It would provide a positive vision for the country on which to focus after the war. It would employ people who would be turned loose from the military after the War. It would unite California, with its gold and natural wealth, with the rest of the country.

Lincoln was able to turn a crisis into an opportunity. The timing for undertaking this task was poor at first glance, since the resources of the country were being consumed by the war. However, the railroad was approved, with government money used to pay the firms which constructed the railroad and to provide land for these firms. This type of government funding would not have been possible before the Civil War because, at that time, the government role in the economy was seen by most as extremely limited. The power given the government during the war gave the government a onetime opportunity. Had the decision been postponed until after the War, it would probably have been delayed indefinitely as the nation attempted to rebuild.

Modern Examples

The manufacturing firm project was progressing well when the crisis struck. A manager picked up a rumor that a competitor was putting in more advanced automation than they were. The competitor was implementing a larger network with more functions. Within hours, a crisis atmosphere pervaded the company. When management held a meeting, the agenda should have included an assessment of the competitor's status and what steps to take to deal with the competitor. Instead, several managers who were enemies of the project immediately began to press for a review of the project and determination as to whether it should be killed off. The challenge was to turn this around. The project leader handed out a short project plan he had developed for the specific crisis. The plan contained material that applied generally to a major crisis. The first set of tasks dealt with verifying what the competitor was doing and the facts. Several tasks to be carried on at the same time concerned determining the potential impact of the competitor's move. The next set of tasks identified what could be done to accelerate their own project.

Lessons learned here are that the project leader should take the lead in analyzing a crisis, and that it is useful to assume the worst in a crisis.

For the construction firm, the pending success of the project spelled disaster to the project schedulers whose jobs would be eliminated or changed. They went to line managers they had worked with and solicited support for an attempt to scuttle the project. The schedulers fed criticisms based on half-truths to the managers. A key attack was that the autonomy of the divisions and project was to be sacrificed by the common project. The attack was launched in a general high-level management meeting. The general manager for the project took the heat and then called in the project manager after the meeting.

Several options were available: confrontation, behind-the-scenes efforts with individual managers, splitting the ranks of the schedulers, or moving the project into a low-profile mode. The lesson learned is to consider a wide range of alternatives.

In order to implement the new product management approach at the consumer products company, a product was selected as a pilot case. Unfortunately, the product selected was neither typical nor simple. The product turned out to be complex in terms of manufacturing, distribution, and marketing. The market segment that would buy the product was still fuzzy. There was also a patent problem. People began to observe and sense the problems. The challenge was to turn the crisis away from the project. The answer was to expand the project to other products and use the problem product as a pilot to show that the new approach could fit the most difficult product.

Management at the manufacturing firm perceived a crisis when they felt that a competitor would have a network and software available first. Management directed that the project be accelerated without obtaining market intelligence. Later, management found that the competitor was not far ahead and was building a smaller network. Management slowed the project and pulled resources. The net effect was harmful to the project. The decisions made resulted in reduced morale and delay in the project schedule, due to failure to plan. The management had failed in a time of crisis.

At the consumer products firm, the product to which the project was attached was atypical and had its own unique complications and projects. The major issue was what to do with a project linked to only one product. A two-pronged strategy was formulated. First, the project would not abandon the product. Instead, the product would be employed as a test to determine the extent of flexibility of the project's approach to a wide range of products. Second, work would begin on applying the project management methods and templates to several other products. This made sense, since considerable progress had been made and it was an appropriate time to test out the project management approach on additional products.

GUDELINES

- **Don't jump to conclusions on either symptoms or causes.**
  Unless a true emergency exists, do not take any precipitate action. Instead, show that you are working on the issue.

- **Ask yourself if there is benefit in forcing a crisis.**

  When faced with a crisis, you have a trump card to play—the timing and posturing of the crisis. You can position the crisis and select the timing of unveiling the crisis and potential solution. If you have been unable to get attention, but you retain management support, consider the cost and benefits of forcing the crisis.

- **Work out a strategy for change in a crisis situation.**

  In project evolution you introduce changes to projects gradually. This can be very smooth at first. However, as time goes on and the number of changes increases, team members experience increasing anxiety and uncertainty.

  In a revolutionary change to a project, a major change is made. This is followed by relative calm as the changes settle in.

  Decide in a crisis whether to take an evolutionary or revolutionary approach.

- **Ask yourself who benefits from how an important issue is resolved.**

  Issues can be addressed, shelved, or made to disappear. They can disappear if one changes the assumptions, purpose, or scope underlying a project. Once an issue is resolved, winners and losers emerge. This is especially true for political issues. Analysis of this outcome can assist you in seeing what is behind the issue and what positions people are taking with respect to the issue.

- **Be cautious when confronted with a critical issue.**

  With a critical issue, the tendency is to plunge in and address the issue. However, first review how the issue surfaced and evolved into being critical. What were the main events along the way? This is important because it helps to frame the issue and give perspective.

- **Draw analogies in project situations to calm temporary project crises.**

  One value of having experienced people in a project is that they can provide perspective and experience when dealing with an issue or a crisis. They can recall a similar event in a previous project. Use these stories to calm people during a crisis.

- **Look for different interpretations of a crisis.**

  When people present you with an issue, they sometimes disguise it by emphasizing the action they desire. Alternatively, they may offer only one view of the crisis. As you have seen, many interpretations are possible. Dig deeper when someone hands you a situation.

- **Learn to enjoy dealing with the unexpected.**

  One reason that people enjoy projects is that projects offer the unexpected. Projects tend to involve change. Change typically involves issues. Issues lead to the unexpected. If you don't enjoy projects, it may be because of having to deal with the unexpected. Learn to enjoy this aspect of project management, rather than dreading it.

- **Play out or simulate what would happen if the immediate crisis were resolved. Ask what new crisis might take its place.**

  Do you have the right interpretation of the crisis? Would the solutions you are considering fit? To help answer these questions, jump ahead mentally and assume that the crisis has been handled. What do you think is the next issue that will become critical and, perhaps, a crisis? Does a link exist between the two issues?

- **Follow up on decisions with actions to maintain management credibility.**

  Managers should consider both decisions and actions together. Through understanding and seeing the actions, you can better evaluate the decisions.

- **Correlate actions with decisions.**

  To evaluate this, make a table with the rows being the decisions and the columns being the actions. In the table write a paragraph on how the action supports the decision. An action may support many decisions, and many actions may apply to a given decision.

- **Compromise to bring later rewards.**

  When considering decisions, maintain an attitude of conciliation and compromise. When you move to actions, you have less flexibility. Decisions have many shades of gray; actions tend to be more black and white. When considering decisions, you can also trade off with opponents on the decisions vs. the actions. If you have to concede points on the actions, you can eventually recover if the actions prove to be inadequate.

- **If you fail to obtain a decision or you disagree with a decision, pause and let the situation alone.**

  Let things cool off. Take time to gain new perspective. Later, you can resurrect parts of issues.

- **Focus on achieving success.**

  This chapter dealt with the termination of a project. If a project is not terminated but is close to failure, you may overcome failure. However, overcoming failure is not the same as achieving success. When you overcome failure, you naturally feel that you have accomplished something. In fact, you have depleted your energy dealing with a problem project.

- **When you stop a project, give a viable explanation for your actions.**

  Stopping a project without explanation fuels rumors. While you don't have to go into detail, cover the major reasons for the decision. Stress the changes that have occurred since the project was started that necessitated your decision.

  Early project decisions are sometimes made in haste and based on external pressures— only to be reversed later.

- **Be cautious in pushing for a decision to terminate a project while you are still in the early stages of the project.**

  This is a period in which the project was recently approved with a specific purpose and scope. A crisis, or at least a major issue, occurs. A manager may start to question the project itself. The logic might be that the project must be poorly conceived if it cannot get a proper start. However, with time, the project may take off and become successful.

- **Avoid placing blame for a failed project.**

  If a project is failing and about to be terminated, no one needs to go down with the ship. Placing blame is not appropriate. If you do blame specific managers, move to have them leave the company. By placing the blame, you may have doomed them while they are at the firm.

- **Frequently ask, "If the project were stopped today, what would happen?"**

  Ask this even if the project is doing fine. Pausing to ask this helps put the project in perspective in terms of its importance to the organization and business processes. It also helps to indicate what is not important in the big picture.

- **Delay a decision in a project to provide time for perspective.**

  This strategy is beneficial for several reasons. The first is that, if the time is not right for a management decision, you will spend energy on the decision but get nowhere. Also, while the decision may be clear, the follow-up actions may still be undefined.

- **Put a project on hold to provide time for perspective.**

  When you place a project on hold, you can still continue work at a low level. You are no longer caught up with the events, so you have time for perspective. Also, a pause gives project team members time to assess how they are doing in their work and what small changes they might make.

STATUS CHECK

- How does your organization identify a crisis? Do you have an organized approach for escalation of a situation? Or do you handle crises on a case-by-case basis?
- Who determines if there is a crisis? What is the role of management? What is the role of the project leader?
- Does each crisis have a learning curve? Do you make an effort to analyze crises after the fact and extract lessons learned?
- Are the decisions that result from a crisis consistent with actions taken? Do actions taken flow logically from decisions?

- Is a method for tracking decisions and actions in place?
- Is a conscious decision process used to relate actions, follow up, and decisions concerning a crisis?

ACTION ITEMS

1. Look back at a crisis in a project (in the past or present, or one about which you have information). Ask yourself the following questions:
   - When was the issue first identified?
   - When did the issue become critical?
   - What were the symptoms that were visible at the time?
   - Looking back, what symptoms should have been visible?
   - What were the causes behind the symptoms?
   - How well trained was the project manager to deal with the crisis?
   - Was management prepared to address the crisis?
   - Could better preparation have been made for the crisis?

2. In a current project, try to identify at least three potential major issues that are either present now or are possible in the future. For each of those, answer the following questions:
   - When was the issue first identified?
   - How much importance do management and the project team associate with the issue?
   - What has been the rate of decay of the situation?
   - What symptoms are currently evident?
   - What are potential causes for the issue?
   - What will cause the issue to become a crisis?

3. Select any three projects that are or were in a state of crisis. For each one, try to determine when the crisis began, what actions were taken, and what results occurred.

4. For these same examples, try to estimate if people had thought through the decisions and actions. What was the elapsed time between the time of the decision and the visibility of the ac-

tions? What additional unannounced and unplanned actions occurred?

5. Assess the side effects of the decisions made by the organization with respect to the project. What was the impact on other projects?

6. Consider any project that was terminated. What has changed with respect to the management of current projects? Do you see any evidence of lessons learned?

CHAPTER 19

100 PROJECT MANAGEMENT PROBLEMS AND WHAT TO DO

CONTENTS

19

100 PROJECT MANAGEMENT
PROBLEMS AND WHAT TO DO

INTRODUCTION

Successful project management is being able to handle issues and problems as well as proceeding proactively to achieve results. In this chapter, 100 issues that might arise in a project are presented.

For each problem, we will consider the following:

- Why?
- Impact
- Action/response
- Prevention

PURPOSE AND SCOPE

The purpose is to identify and address potential problems in the following categories:

- **External Factors** These include other projects, the industry, competition, and factors beyond your detailed knowledge and control.

- **Management** This includes both management of the project and the team.

- **The Project and Project Plan** This includes all facets of planning, analysis, and direction.

- **The Project Team and Resources** This covers the management and coordination of all project resources, including the project team.

- **Methods and Tools**　This encompasses methods and tools of project management and of the project itself.
- **The Project Work**　This is the direct management and oversight of the work.

APPROACH

We will now present the problems, their impact, how to handle the problems, and suggestions for preventing these problems.

External Factors

1. Delays in related projects impact tasks.

   Why? Dependencies between projects were not evident. No one was in charge of pulling together all of the projects.

   Impact: Tasks slip. The morale of the project team may be lowered.

   Action/response: Project management should stop and define relationships between projects.

   Prevention: Implement overall project coordination.

2. No one set up or thought about interfaces between projects.

   Why? Ego and the desire to be self-sufficient contribute to this problem.

   Impact: The project keeps running into walls generated by interface problems.

   Action/response: Define interfaces.

   Prevention: Identify all relevant projects and systems and hold planning sessions with the team on the interfaces.

3. Too much of the project depends on external organizations that cannot be controlled.

   Why? Since these organizations are not under the control of the project team, team members feel powerless to implement plans to assist the project team in dealing with the external organizations.

Impact: The project team and leader feel helpless to remedy the situation.

Action/response: For each organization, consider what its self-interest is; establish contacts to work with each organization on that basis.

Prevention: Make a chart to map each organization in terms of the effect of the project on its self-interest.

4. The facilities of the project are not conducive to work.

   Why? People don't think about the facilities until they start working in them.

   Impact: Loss of productivity.

   Action/response: Recognize that the facilities are a problem and make a list of relevant issues. Then deal with the issues.

   Prevention: Evaluate the facilities before the project begins and, if necessary, formulate a plan for redesigning the facilities.

5. Competitors working on similar projects are ahead of your project.

   Why? Beyond having more resources or a head start, the reason for this may also lie in the competitor's approach to the project.

   Impact: Pressure on your project team may increase.

   Action/response: Try to determine how the competitor's project is organized and what the competitor's technical approach is.

   Prevention: While not preventable in every case, you can survey what other firms are doing in the area of the project at the start of your project.

6. No one considered the impact of external factors at the start of the project.

   Why? This occurs in organizations that are internally focused.

   Impact: You may face one surprise or issue after another.

   Action/response: Refocus the project to deal with the impact of external factors on the project.

   Prevention: From the beginning of the project, consider the range of external factors and what their potential impact might be.

7. No systematic method is in place for identifying and addressing external factors.

Why? The team fails to address the need for a systematic approach to project management.

Impact: Control of the schedules and work is taken over by events. Project work effectiveness is impacted.

Action/response: Demonstrate to management the impact of not considering the external factors.

Prevention: Make the assessing of outside impact a natural part of the process. Consider selecting a project and using this as a model of analysis.

Management

8. Management does not make the effort to obtain decisions.

Why? People may assume that decisions will be made automatically on a timely basis.

Impact: Project work may continue based on a set of assumptions on a management decision. The decision may eventually be different from what was assumed, disrupting the project.

Action/response: Gently press for a decision. Indicate what direction the project will take until the decision is made.

Prevention: Prepare managers for the pending decision required. Try to get acceptance of an informal decision in the interim.

9. Decisions are being made ad hoc and not on an organized basis.

Why? Management could be attempting to demonstrate that it is action-oriented. Any problem gets an immediate decision.

Impact: The decisions may be inconsistent.

Action/response: Try to package a set of issues to obtain more systematic decisions. Point out alternative decisions and their impact.

Prevention: Take a systematic approach to identifying and analyzing issues.

10. Consistent project controls are lacking.

    Why? The organization may never have taken the time to establish project controls in terms of resource use and budgeting.

    Impact: Each project ends up having a slightly different set of controls based in part on the style of the participants.

    Action/response: Select a specific project and use this to establish controls.

    Prevention: Choose a specific project and use it as a model for other projects.

11. Projects receive uneven management attention.

    Why? This is almost inevitable, given the differences between projects.

    Impact: This depends on the basis of the inconsistency. If it is due to style and management interest, this is more of a problem. Projects that deserve more attention may languish.

    Action/response: Take a project that should get more attention and attempt to elevate it in terms of visibility.

    Prevention: Develop an approach for prioritizing project reviews.

12. Standard project reporting is lacking.

    Why? The organization gradually moved into project management without formulating a standardized approach to project reporting.

    Impact: The projects cannot be compared. This means less opportunity to extract lessons learned and apply these to other projects.

    Action/response: Use several projects as models and build a standardized reporting structure.

    Prevention: Develop a project summary of all projects. This can serve as a top-down approach for handling project reporting.

13. Some in the organization oppose the project—a factor not addressed by management.

Why? Upper management assumes that their endorsement is sufficient to make the organizations fall into line.

Impact: Management and the project team are thinking one thing; others in the organization are working from another set of assumptions. Problems will arise in reviews and in getting resources.

Action/response: Attempt to get everyone on board. Use direct contact and market to these groups.

Prevention: Attempt to align the organization with the project by explaining the roles of the various organizations and the benefits of the project, emphasizing the aspects that will appeal to the self interests of the different groups.

14. Actions that follow from decisions are not consistent with the decisions.

    Why? No one thinks ahead to how the decisions should be implemented. Faced with the actions, management reverses decisions.

    Impact: The result may be loss of credibility in the decisions.

    Action/response: Begin immediately to identify actions and have them approved with the decision.

    Prevention: Package the actions with the decisions.

15. With the attention on a few large projects, management ignores key small projects.

    Why? People can easily be swept up in the large projects. Everything else is treated as if it could be addressed later.

    Impact: Issues and decisions related to smaller projects languish.

    Action/response: Bundle the small projects together and present them as a package to get more attention.

    Prevention: Take an overall project view that increases the visibility of each of the small projects.

16. Follow-up is neglected after decisions are made.

    Why? Decisions are made. Actions are taken. Management may not go on to the next decision, assuming that no follow-up is needed.

    Impact: Actions and decisions become ineffective.

    Action/response: Make follow-up part of the actions.

    Prevention: Include measurement of the actions in the decisions process.

17. The project has no organized forum to address issues.

    Why? Management addresses each project individually. This leads to an inconsistent approach to issues.

    Impact: This creates the potential for conflicting decisions.

    Action/response: Start categorizing the issues to show patterns.

    Prevention: On a proactive basis, develop a consistent approach for the issues.

18. Managers are dabbling in the project.

    Why? Managers may be attracted to a specific project due to the fascinating technical aspects or the financial importance of the project.

    Impact: The project may become micromanaged, holding up progress.

    Action/response: Move to create summaries of the project and to reduce visibility of the project.

    Prevention: At the start, define management's role in the project.

19. Management doesn't hold the project team accountable.

    Why? Managers take on too much responsibility and let the team off the hook.

    Impact: The project team loses a sense of accountability.

    Action/response: Instill more accountability with the team at the task level.

    Prevention: Define accountability for the team and management at the start of the project.

20. Significant top management change occurs.

Why? This is a normal course of events.

Impact: New leadership may slow progress on the project while learning the job.

Action/response: Try to work with the new management immediately by getting them on board the project.

Prevention: Adopt the approach that the project depends on a management team, rather than on any one person, and that the organization has a vested self-interest in the project.

21. The need for the project has evaporated.

Why? The organization mission or roles may have changed.

Impact: The project will be canceled or redirected.

Action/response: Consider a project review that may result in terminating the project.

Prevention: Keep the project purpose and scope in mind. Review these on a regular basis and update them if needed.

22. The project is unsuccessful in competing for resources.

Why? Due to political factors or the excitement generated by another project, your project suffers.

Impact: The project will have to get by with fewer resources.

Action/response: Consider restructuring the work to a lower resource level.

Prevention: Begin with a small number of resources and don't count on many additional resources.

23. The project assumes a political life of its own.

Why? While the project team makes some progress, the major milestones are never reached. Since management does not intervene, the project team assumes that everything is fine. People may lack the will to kill the project.

Impact: Resources are wasted if the project continues when it should have been terminated.

Action/response: Instead of confronting team members, move the resources away from the project and let the project starve.

Prevention: The best approach involves labeling the project as political at the start and then imposing controls on the project.

24. Milestones are not reviewed.

Why? A regular process of review may not have been instituted. Review may be seen as indicating a lack of trust in the team.

Impact: The problem will surface when it becomes apparent that the work is not satisfactory. The entire project may be questioned.

Action/response: If a review is not imposed by management, define your own review process.

Prevention: Define a formal review process.

25. Management is too formal and emphasizes status as opposed to content and issues.

Why? This often reflects the current management style.

Impact: Time for dealing with projects is consumed by concerns with status. Issues and substance are ignored. Later, crises arise that consume management attention.

Action/response: Begin to present issues as a part of addressing status.

Prevention: Try to institute a process in which status and issues are dealt with separately.

26. Ability or desire to pull all of the projects together for analysis does not exist.

Why? Perhaps no one thought of doing this. Also, the projects may be in different formats and structures, making summaries more important.

Impact: Managers lack an overall sense of how well projects are being managed.

Action/response: Create a summary project plan. This is a rollup of the active projects into one project. Then you will be able to see how all projects work together.

Prevention: Establish a summary reporting process.

27. An organizational process for issues is missing.

Why? Management may never have been comfortable with project management. Issues from projects are treated the same as those from line organizations.

Impact: Without a process, the issues are addressed one at a time and not systematically.

Action/response: Propose developing a standard process.

Prevention: Develop the issues process, as discussed in Chapter 18.

28. Management overreacts to issues.

Why? This may be a natural occurrence. The problem arises when management begins to overreact to many issues.

Impact: If the team members perceive the overreaction, they become reluctant to bring issues forward. The issues tend to sit until there is a crisis. This in turn feeds the overreaction.

Action/response: Attempt to provide an overall perspective of each project in a summary mode. This can help managers step back from the detail.

Prevention: Implement a structured approach for issue management that allows for escalation.

29. Misunderstandings about the project are not corrected.

Why? With project communications extending over months, misunderstandings are likely to occur.

Impact: Misunderstandings can make enemies of the project. Misconceptions can fester. These can then affect decisions related to the project.

Action/response: Tackle misunderstandings head-on. Don't wait.

Prevention: Establish regular communications with managers related to the project.

The Project and Project Plan

30. A baseline plan does not exist.

Why? A firm does not get in the habit of measuring actual vs. planned results.

Impact: Without a baseline plan, the plan itself can shift and decay.

Action/response: Set baseline plans for all active projects. Do not wait for a new project to set the baseline. Then compare the actual with the planned or budgeted results.

Prevention: Implement a planning process that includes the baseline vs. actual comparison.

31. Resource consumption is too variable.

    Why? This occurs when the demand for resources varies with the types of tasks that are active.

    Impact: Management of the work is more difficult, since the project manager must constantly adjust to the workload and staff members.

    Action/response: The variability may be planned, so adjust the approach for managing the resources.

    Prevention: Through planning, try to move task schedules.

32. Project meetings lack focus.

    Why? This occurs when no standard method for handling meetings is in place.

    Impact: Issues remain unresolved.

    Action/response: Step in and impose a standard format for meetings and then control the meetings.

    Prevention: Establish standard meeting formats with sample agendas.

33. A lack of awareness of history of the project exists.

    Why? People treat all issues as new.

    Impact: The experience gleaned from working on past issues is not applied to the present.

    Action/response: Someone should build a set of lessons learned from the project. This can serve as the basis for a history.

Prevention: Build a history of the project from the start around issues and lessons learned.

34. The project is being treated too routinely, like past projects.

    Why? This occurs when people are too confident due to success on past projects.

    Impact: When you treat a project as routine, people don't take the issues seriously. This can lead to a crisis.

    Action/response: Try to put a spark of life in the project by emphasizing what is new and different about this project.

    Prevention: Clearly define what is new and what can be learned from past projects.

35. The project plan lacks key resources.

    Why? The person creating the plan left out important resources.

    Impact: Resources must be requested on an ad hoc, disorganized basis.

    Action/response: Update the project plan with all of the resources that you require.

    Prevention: Assign people to review the list of resources that are in the project.

36. Too much time elapsed between milestones.

    Why? This can be the result of inexperience, or lack of planning and review of the plan.

    Impact: With substantial elapsed time between milestones, one begins to wonder what is going on in the project. What is the status and progress?

    Action/response: Find out what is going on. Then define more milestones to reduce the gap between the existing milestones.

    Prevention: Attempt to reach a milestone every three to four weeks. Define interim milestones every two weeks.

37. The project gets off to a slow start.

    Why? The project may have no early actions. The tasks may be related to purchasing and hiring.

Impact: The project languishes, but management expectations are still in place. The risk is that progress will not resume.

Action/response: Begin on parallel tasks. Demonstrate activity.

Prevention: With a standard template and detailed tasks, many initial, parallel tasks should be scheduled.

38. The plan is not kept up-to-date.

    Why? People are busy doing work and don't pay attention to the plan.

    Impact: Without updates, the plan begins to lack credibility. Things begin to fall apart when management notices that the project plan is out-of-date.

    Action/response: Set a baseline schedule. Impose regular updating once a week.

    Prevention: Require that an overall plan of all projects be created. All plans and projects should follow a regular update process.

39. The tasks are more sequential than was planned.

    Why? This is likely due to poor organization of the work and plan. This may be due to lack of experience.

    Impact: The schedule will now slip.

    Action/response: Sit down and start analyzing the plan to set new dependencies. Then review the plan.

    Prevention: The best prevention is to have a review by the people who will be doing the work.

40. The plan is too detailed.

    Why? Some first-time project managers want to do the work "right" and define a great many tasks down to the level of a one-day or two-day duration.

    Impact: Such a plan cannot be updated or changed without great effort. This may lead to the plan not being updated at all.

    Action/response: Cut back the detail in a new schedule. Make the shortest task at least two weeks duration.

Prevention: Use a template and instruct managers on rules relating to detail.

41. The purpose of the project was not clear at the start.

    Why? The purpose of the project was not reviewed along with the scope.

    Impact: Effort will be wasted and people will be working on the wrong tasks.

    Action/response: Redefine the purpose to management and then revise the schedule to be aligned with the purpose.

    Prevention: The purpose and scope should be defined first. Then all milestones should be determined and evaluated in light of the purpose.

42. The scope of the project is expanding.

    Why? This may be due to a lack of management controls. The phenomenon of scope creep is common. This occurs with promising projects in which people want to tack on just one more milestone.

    Impact: The project schedule slips and the budget balloons.

    Action/response: Separate out the items that should be deferred. With the scope redefined, return to the plan and revise this to fit the scope.

    Prevention: Impose rules on the start of the project relating to any out-of-scope issues.

43. Despite an expanding scope, no budget adjustments or resource additions are made.

    Why? The people who approved the scope change are not often the ones who approve budget and resource additions. Thus, the resources and budget don't fit with the scope.

    Impact: The project will grow out of control.

    Action/response: After reviewing the scope, go to management with the choice of additional resources or reduced scope.

    Prevention: Impose a rule that all scope changes of a certain level and above must have resource and budget review.

44. People are unwilling to change the structure of the project.

Why? People may feel comfortable with the project structure. They are unwilling to change, fearing, perhaps, that this will cause a management review.

Impact: The structure of the project does not reflect the reality of the work. Project reviews will be more artificial and will require more effort.

Action/response: Revisit the structure of the plan and make revisions.

Prevention: Make a policy to review the plan overall at three-month or six-month intervals and make structure changes.

45. The plan is not synchronized with the actual results.

Why? The plan may not have been updated. The updates to the plan may not be complete. The additional out-of-scope and rework tasks are not added.

Impact: People begin to ignore the plan since it is not realistic.

Action/response: Conduct a review of the status of the actual work. Then go to the plan and update this to fit the reality.

Prevention: Conduct a reality check at regular intervals.

46. Issues are being addressed one at a time rather than in groups.

Why? No systematic, organized approach to address issues is in place.

Impact: The resolution of successive issues may be contradictory. Issues may sit unresolved, while time is spent on analysis of past decisions.

Action/response: All outstanding issues should be grouped and then organized for review.

Prevention: A process for identifying and dealing with issues should be included in the plan.

47. The project plan is changing too much.

Why? The project leader may be attempting to have the plan reflect every day events.

Impact: With so many changes, the credibility of the plan is in doubt. This can occur when several versions of the plan are in circulation at one time, leading to more confusion.

Action/response: Procedures for updating and changing the plan should be established.

Prevention: Guidelines for changing the plan and for employing standard templates should be established.

48. The level of effort required is consistently underestimated.

Why? This can be due to inexperience or due to unrealistic target dates.

Impact: The plan and schedule will immediately slip and continue to do so. The plan will lose its meaning.

Action/response: All changes and updates to the plan should be stopped while people realistically estimate duration and effort.

Prevention: The best prevention is to have a detailed project review at the start and additional reviews periodically.

49. Significant tasks are missing.

Why? The project team or leader may never have worked on this type of project. Also, this can occur when substantial parts of the project are not within the control of the team.

Impact: The schedule will stretch out as these tasks appear. If the tasks are not added to the plan, the plan will not reflect the work.

Action/response: Review the project structure with the additional past, present, and future work added.

Prevention: Schedule an extensive review of the initial project plan prior to setting the baseline schedule.

50. The project has too many detailed tasks.

Why? The project team may supply detailed tasks to the project leader, who adds them to the plan.

Impact: Updating the plan may be difficult. Reviewing and reading the plan for understanding will be more complex.

Action/response: Create a summary schedule from the existing plan and then selectively add detail.

Prevention: Use a template together with guidelines on adding more detail to the work.

51. The wrong tasks are deleted in trying to meet the schedule.

    Why? The project leader may panic and try to cut tasks just to meet the deadline.

    Impact: Slippage will occur as the now missing tasks impose their impact.

    Action/response: Revisit the structure of the plan and expand the tasks. Review the new schedule with management.

    Prevention: Draw up guidelines for revising a project plan and for task deletion.

52. The number and frequency of milestones are insufficient.

    Why? Inexperience is one cause. Optimism or relying on strategies from a past successful project may also make people careless.

    Impact: The project will be more difficult to track.

    Action/response: Revisit the schedule and identify additional milestones.

    Prevention: If a standard template is used, this should have sufficient milestones.

53. Integration of the work in the project suffers from lack of attention.

    Why? Integration is one of the more difficult task areas, since it is technically complex and involves many people and organizations. Hence, the temptation is to simplify the integration tasks.

    Impact: When integration tasks are being worked on, the schedule is likely to slip due to missing tasks.

    Action/response: Consider setting up a separate project for integration only. Then combine this with the basic project.

    Prevention: Consider integration a key area when reviewing the schedule prior to setting the baseline schedule.

54. The project is going on too long.

    Why? This is the nature of the work.

Impact: The project may begin to drift out of control. People may get caught up in the process of the project as opposed to the work.

Action/response: Consider dividing the project into a number of separate parts.

Prevention: At the start, divide the project into subprojects, each of which has separate milestones.

55. The plan is too informal.

Why? People may never have prepared a detailed project plan. They create a general plan that appears to be acceptable.

Impact: The milestones and progress are difficult to evaluate and track.

Action/response: Impose formal project management and planning on the project.

Prevention: The best prevention is to have standard templates and guidelines.

56. The project plan should be changed significantly, but it is not.

Why? The project leader may be reluctant to make changes that might disrupt the project.

Impact: The project plan does not reflect what is going on in the project.

Action/response: Make a general project change and then impose standard reviews.

Prevention: Define guidelines for changing the schedule.

57. The project plan is not adaptive to results or events.

Why? This may be due to a template or schedule that is too rigid. Too many dependencies result.

Impact: The plan may end up being obsolete.

Action/response: Consider developing a new plan that is more flexible with summary tasks and fewer dependencies.

Prevention: The use of a template combined with tests of the template and lessons learned should make the plan more adaptive.

The Project Team and Resources

58. The project leader lacks personnel skills.

    Why? The project leader may be assigned to the project based on availability, rather than necessary skills.

    Impact: The project team and project suffer while the leader goes through a learning curve.

    Action/response: In extreme cases, the project leader may be replaced. A less severe approach is to assign a mentor to the project leader.

    Prevention: Consider assigning a senior project leader to oversee the project at the start.

59. No organized approach for sharing of information has been formulated.

    Why? People may mistakenly assume that information will be shared.

    Impact: People may be working on the project in a vacuum. Without information sharing there will be less coordination.

    Action/response: Consider having meetings in which information is shared relative to issues.

    Prevention: Establish guidelines to support lessons learned and sharing of information.

60. Too much bureaucracy rules in the management of the project.

    Why? This may be due to the style of management or to the style of the project leader.

    Impact: Bureaucracy puts measuring results in the back seat. The project looks good on paper. In reality, it may be falling apart.

    Action/response: Without replacing the project manager, move the emphasis of the project toward issues rather than status. Reduce the overhead tasks.

    Prevention: Guidelines for the project leader should be in place.

61. People are spread too thin in the project.

    Why? The project may depend on a few key people who are used in many tasks.

Impact: The plan may start to slip as the team members become stressed and overworked.

Action/response: Consider reassigning some of the tasks or dividing tasks up so that more junior employees can do some of the work.

Prevention: In planning for the project, assign resources to the project, then filter the project to see how overcommitted people are.

62. The project leader fails to delegate to the team.

    Why? The project leader may be attempting to do too much of the work himself or herself.

    Impact: The project leader becomes a bottleneck and the team members get more frustrated.

    Action/response: Reassign some of the tasks to team members.

    Prevention: Clearly define and review the role of the project leader.

63. Regular flow of end products from contractors is not happening.

    Why? This may be due to poor planning on the part of the project leader, as well as to inexperience in dealing with contractors.

    Impact: Project control can shift from the project leader to the contractor.

    Action/response: Sit down with the contractor and define a new project plan for work with this contractor. Then merge this with the current plan.

    Prevention: In defining milestones, make sure that the work of the contractors can be tracked.

64. Team assignments are being changed too often.

    Why? Due to the pressures of the work and outside influence, the team may be shifted frequently.

    Impact: The team may lose faith in the project leader and feel that the project is in chaos.

    Action/response: Revisit the project plan and assign areas of tasks to team members.

Prevention: Make team assignments to areas of tasks, not individual tasks. Change team assignments on an infrequent basis.

65. People are kept on the team too long.

Why? Personal relationships may have been established so that the manager is reluctant to reassign the people.

Impact: Morale may suffer as some team members feel that others are dragging down the project.

Action/response: Review all roles and responsibilities on the team. Move to reassign or remove some team members.

Prevention: Explain clearly to each team member at the start when he or she will enter and leave the project team. Stick to the schedule.

66. The project manager is spending too much time in project administration.

Why? This can occur in a large project with many administrative tasks. This can occur in smaller projects when the project leader prefers administrative tasks.

Impact: The management side of the project leadership will suffer.

Action/response: The project leader should be counseled and guided to spend more time on management.

Prevention: The role and duties of the project leader should be clearly defined at the start.

67. Miscommunications with line management occur.

Why? People spend too much time inside the project. The project leader is internally focused.

Impact: Miscommunications can lead to misunderstandings, which in turn lead to resistance and hostility.

Action/response: Open up the lines of communications with line management by discussing issues and status of the project.

Prevention: The approach for working with line managers should be defined at the start of the project.

68. The project manager lacks management skills, and no team member fills the vacuum.

    Why? While people realize the problem, they let this ride rather than cause additional problems by interfering.

    Impact: Management issues remain unresolved. The project team senses a lack of management.

    Action/response: Consider reassigning some duties to senior team members or adding a senior project leader to address issues.

    Prevention: The lack of management skills known up front can be accommodated in planning and staffing.

69. The project suffers from uneven distribution of skills and knowledge.

    Why? This is a natural occurrence in medium and large projects.

    Impact: The project may fall back and depend on the key people. The talents of the junior team members are wasted.

    Action/response: Consider subteams within the project team.

    Prevention: Pair junior team members with senior members.

70. Team members put out false information.

    Why? The team members may be unhappy in the project. They may not like the tasks to which they are assigned. They attempt to undermine the project.

    Impact: This creates more work and requires the project leader to exercise damage control.

    Action/response: Step in and counter the misinformation. Also, have a one-on-one meeting with the team member to exert control.

    Prevention: Establish rules of project conduct at the start of the project.

71. Too many people are doing project management.

    Why? This may be due to a vacuum of leadership which causes people on the team to step into the void.

Impact: Management becomes diffused. Issues are much more difficult to resolve.

Action/response: Revisit the management role within the project.

Prevention: Carefully assign all management tasks and responsibilities at the start of the project.

72. Team members fail to learn lessons.

    Why? Perhaps no process was instituted for collecting and using lessons learned. People are overly focused on the present.

    Impact: People repeat variations of the same mistakes.

    Action/response: Institute a project review to gather lessons learned so far in the project. Update these on a regular basis.

    Prevention: Guidelines at the start of the project should stress the importance of and the approach for benefiting from lessons learned.

73. People are spread too thin between high priority projects.

    Why? Different projects may require the same set of skills and knowledge, placing stress and demand on key people.

    Impact: The worst case sometimes occurs—all projects suffer.

    Action/response: Management from all projects should get together and set priorities.

    Prevention: Draw up an overall summary project plan that links all projects. That will allow you to see where resources are scarce and assist you in allocation.

74. The style of the team is affecting the project too much.

    Why? This can be due to a lack of strong leadership on the part of the project manager.

    Impact: The project is not under consistent control.

    Action/response: Take control of the team and adopt more formal methods.

    Prevention: Set the tone and the level of formalism at the start.

75. Boundaries at the edge of the project (indicating scope) are breaking down.

Why? People in the project team are too helpful —they perform more work than that which falls within the scope of the project.

Impact: The scope is expanding, but the resources and schedule are not changing.

Action/response: The boundary of the project needs to be firmly drawn and enforced.

Prevention: Establish rules for changing the scope and taking on more work at the start of the project.

76. The anticipated resources fail to appear.

Why? People assigned to help you out when the project began are not available due to priorities of other projects. Similar problems can occur with equipment and facilities.

Impact: The project can slip due to the lack of resources.

Action/response: Line up alternatives.

Prevention: Check availability of resources at least one month prior to when you need them and then check weekly prior to the arrival of the resources.

77. Management communications gradually deteriorate.

Why? This can be due to lack of contact and misconceptions.

Impact: Support for the project may be threatened.

Action/response: Begin to reestablish communications on a regular basis.

Prevention: Establish a communications pattern early in the project to maintain regular, expected contact.

78. Personality clashes lead to poor communications.

Why? This is almost inevitable on projects with long duration.

Impact: Project meetings are strained. There is a lack of coordinated effort on tasks. The schedule and quality may suffer as a result.

Action/response: Get the people together. Indicate that you are not trying to force them to get along or be friends but, for the sake of the project, you expect some minimum level of communication.

Prevention: Track communications early in the project and try to head off problems.

79. The subcontractor team is not managed well.

    Why? This can occur when the leader does not provide supervision.

    Impact: The lack of organization and management among the subcontractors may impact the project negatively.

    Action/response: Contact the manager of the subcontractor and negotiate rules.

    Prevention: Prior to the start of work, establish policies on how the subcontractor staff will be managed.

80. Too much of the project is being done by one person.

    Why? In many projects, one particular person tries to be helpful. This person takes on more and more work.

    Impact: The first impact may be to delay the project. Another effect is to create a bottleneck. The project may suffer from that person being a single source of information. No backup is prepared to take over if this person becomes ill or leaves.

    Action/response: Reassign the work and distribute it to more than one person to prevent recurrence of the problem.

    Prevention: Establish clear boundaries at the beginning of the project concerning what people are to work on.

81. Poor performance by some drags down performance by all.

    Why? Uneven performance is frequent on large projects involving many people.

    Impact: The other team members see that nonperformance or poor performance is not addressed. They may then have a tendency to slack off in their work.

    Action/response: Performance of the entire team should be measured. At the same time you are addressing poor performers, praise good performance.

    Prevention: Indicate the policies related to performance review and follow up with statements about performance on a regular basis without mentioning names.

82. People are too relaxed after reaching a key milestone.

    Why? The team may have worked very hard toward achieving a specific milestone. They made it and they were praised. Then they take a break. The break continues too long.

    Impact: Time gained and progress made are now being dissipated.

    Action/response: Begin the work on the next milestone prior to the completion of the current one. Focus initially on smaller level tasks to get the momentum back.

    Prevention: Institute overlapping milestones in terms of the work leading up to the milestones.

Methods and Tools

83. Even though a method or tool has proven to be a failure, it is not discarded or replaced.

    Why? People may not want to acknowledge failure or problems. They may still pay lip service to the method or tool.

    Impact: A method or tool gap is being addressed on an ad hoc basis by each person. This leads to reduced productivity as each person struggles separately.

    Action/response: Announce the replacement method or tool at the same time that you kill off the old method or tool.

    Prevention: Carefully evaluate and select the methods and tools at the beginning of the project.

84. Different people are employing different, incompatible tools.

    Why? If no standards are established concerning the tools at the start of the project, people often employ the tools that they have used before.

    Impact: Incompatible tools mean that a person may have to understand multiple tools that do the same thing. Additional work may result from trying to make one tool work with another.

    Action/response: Review the existing tools and implement a phaseout of some.

Prevention: Identify the tool set at the start of the project. Reinforce proper use of the tools with training.

85. Staff members are resistant to the methods or to the tools.

    Why? If training is not good or not provided, or documentation is poor, staff may get a bad impression of the methods and tools.

    Impact: If staff members are resistant, they may not use the methods or tools properly. They may resist using the methods and tools at all.

    Action/response: Meet with the people and identify why they feel uncomfortable about the methods or tools. Take steps to rectify the situation.

    Prevention: Having announced and reinforced the methods and tools at the start of the project, follow up by monitoring their use.

86. The project suffers from a lack of experts in the methods and tools.

    Why? This may occur if your project is the first in the company to use the method or tool.

    Impact: Productivity falls as people attempt to use the method or tool.

    Action/response: Appoint someone to learn how to use the tool and then train the others.

    Prevention: Identify the future experts in tools at the start of the project.

87. No one took enough time to learn the method or tool.

    Why? The pressures of the schedule precluded any substantial learning.

    Impact: The project team is not able to take advantage of the features of the method or tool.

    Action/response: Prepare basic guidelines for the method or tool.

    Prevention: Add tasks related to learning the method or tool to the project plan and template.

88. Bad habits in using tools and methods are not addressed or are unlearned.

    Why? This can happen because of laziness.

    Impact: Poor tool use reduces productivity and may affect the schedule.

    Action/response: Have people share experiences and lessons learned on how to employ tools.

    Prevention: The use of the tool expert and mentor, along with reinforcement training, can help keep bad habits from forming.

89. No new tools were investigated during the project.

    Why? People were too busy doing the work; even though promising tools were available, they were ignored.

    Impact: The biggest impact is the loss of opportunity. However, there is no way of knowing the size of the impact, since this is a trade-off with conversion and training.

    Action/response: Conduct a tool assessment to see what is available.

    Prevention: Make tool evaluation a task in the project.

90. A new tool is inserted in the project without planning.

    Why? Some manager gets excited about a tool and buys one without accompanying planning and training.

    Impact: No one thought through the learning curve or how people were going to integrate this new tool with other tools.

    Action/response: Stop the use of the tool and assess its role and learning curve.

    Prevention: Implement a planned approach for tool evaluation, selection, and implementation.

91. No one knows whether the methods and tools in the work are really helping.

    Why? No organized approach is taken to measure the project internally.

    Impact: Doubts may arise about the project itself.

Action/response: Institute a review of methods and tools by having the project team share lessons learned. Document these lessons learned and update them.

Prevention: Make measurement of the methods and tools a regular part of the project.

The Project Work

92. The project requires too much rework.

    Why? This can be due to poor quality or lack of training.

    Impact: The schedule will slip and the credibility of the plan will suffer.

    Action/response: Investigate where the rework time needs to be done, then work backward to find out where the problems are.

    Prevention: Keep an eye on quality and review of work. This can translate into an analysis of rework. The rework should also be measured closely.

93. No one is looking for improvements.

    Why? In some projects the actual techniques related to work are frozen in time at the start of the project to achieve better stability.

    Impact: The opportunity for change and the benefits that could result from the changes are lost.

    Action/response: People on the project team should meet to discuss how they do their work.

    Prevention: Incorporate improving the work as a regular task in the schedule.

94. Too many tasks are unplanned.

    Why? This can be due to the nature of the project. This can also be due to the lack of planning at the start.

    Impact: Unplanned tasks make the schedule extend in time and create more resource conflicts.

    Action/response: Start by dealing with the unplanned tasks that occurred in the past. Then expand your work to tasks in the near future.

Prevention: The best prevention is to gather people together to identify possible issues that may arise. Do this at the start of the project. By identifying issues, you can determine related tasks.

95. Quality of the work is an issue.

Why? Management may put too much emphasis on schedule and not enough on quality.

Impact: Poor work leads to rework and extra work, affecting the schedule.

Action/response: Make quality an important part of a review, as well as a point of emphasis during the work.

Prevention: Spell out quality standards at the beginning of the project.

96. The project is hit with compound issues and crises and does not recover.

Why? The project can be a bad luck project. However, this can also occur as a result of attack from enemies of the project.

Impact: The project may grind to a halt.

Action/response: Identify and assess the outstanding issues. Then address the issues.

Prevention: Warn people that crises may come in bunches and that they should be ready to address these through issue management.

97. What has been accomplished is not reviewed.

Why? People may be in a hurry. The review will cause the schedule to slip.

Impact: The project is exposed because of the lack of review. Additional rework may be needed later.

Action/response: The level of review should be set ahead of time. This is a planned trade-off to determine how much to do with available resources.

Prevention: Schedule reviews for all significant milestones. Assess the level of effort needed vs. the risk and exposure.

98. Information on how to evaluate the work is lacking.

    Why? Without preparation for doing a review, the review may accomplish very little.

    Impact: The major impact is that management gets the impression that the review was complete, when it was only superficial.

    Action/response: Distribute packets of material prior to the review.

    Prevention: Include preparation for reviews as tasks in the plan.

99. One issue dominates the work in the project—leading to problems with other tasks.

    Why? This is natural when one issue is dominant financially or politically.

    Impact: Other issues may be lost in the shuffle.

    Action/response: Include the other issues in an issue summary. Allocate time to a group of issues separate from the main issue.

    Prevention: This will occur. You can only mitigate the impact with an organized prioritization of issues.

100. Issues surfacing around the work are not taken to the project leader and team.

    Why? People may be reluctant to bring problems to management.

    Impact: The issues tend to fester. They are unlikely to diminish. The issues then can impact the project.

    Action/response: Spend time getting acquainted with the day-to-day work of the project. Gather up issues and match those with the plan and the issues list.

    Prevention: Initiate more regular communications between the project team and any other workers in the project. Encourage and monitor an open communication process.

STATUS CHECK

- In reviewing the problems and issues listed, in which areas have you seen problems?
- Which problems have you encountered in the past six months?

CD-ROM ITEMS

19-01   Rylande Project Issues

19-02   Issues

The 100 issues in this chapter are given in a spreadsheet.

APPENDIX

SOURCES OF INFORMATION

SOCIETIES AND GROUPS YOU CAN JOIN

Here is a list of some of the organizations involved in project management.

- Association for Computing Machinery
  Web Site: www.acm.org
  1515 Broadway
  New York, NY 10036–5701
  Phone: 800–342–6626
  Fax: 212–944–1315

- The Institute of Electrical and Electronics Engineers, Inc.
  Web Site: www.ieee.org
  3 Park Avenue, 17th Floor
  New York, NY 10016-5997
  Phone: 212-419-7900
  Fax: 212-752-4929

- Institute of Industrial Engineers
  Web Site: www.iienet.org
  25 Technology Park
  Norcross, GA 30092
  Phone: 770–449–0460
  Fax: 770-441-3295

- International Society for Enterprise Engineering
  Web Site: www.iseenet.org

c/o Dennis Murphy
GTE
P.O. Box 223011
Chantilly, VA 20153-3011
Phone: (703) 818-5398

- Project Management Institute
  Web Site: www.pmi.org
  Four Campus Boulevard
  Newtown Square, PA 19073–3299
  Phone: 610–356–4600
  Fax: 610–356–4647

- Reliability Engineering
  Web Site: www.enre.umd.edu
  University of Maryland
  2100 Marie Mount Hall
  College Park, MD 20742

- Software Engineering Institute
  Web Site: www.sei.cmu.edu/about/about.html
  Carnegie Mellon University
  Pittsburgh, PA 15213–3890
  Phone: 412–268–5800

Look for articles, case studies, and books offered by these groups.

SOURCES OF GENERAL PROJECT MANAGEMENT INFORMATION AND EXAMPLES

Articles on projects and project management can be found in many industry-specific magazines related to transportation, construction, government, medicine, banking, insurance, and retailing. You may be able to obtain these through a complimentary subscription.

Project management articles are also found in the following publications:

- *Project Management Journal*   www.pmi.org
- *Project Manager Today*   www.projectnet.co.uk

SOURCES OF PROJECT MANAGEMENT SOFTWARE INFORMATION

- *Datamation*   www.datamation.com
- *Information Week*   www.informationweek.com
- *ComputerWorld*   www.computerworld.com
- *Software Magazine*   www.softwaremag.com
- *CIO Magazine*   www.cio.com/cio

SOURCES ON THE WEB

Enter "project management" into one of the standard search engines and you will find a wealth of information. One specific source is the WWW Project Management Forum at www.pmforum.org.
Information available on the Web includes the following:

- Societies concerned with project management
- Software vendors and products
- Project management consultants
- Online bookstores

GUIDE TO SUCCESSFUL PROJECT MANAGEMENT

INDEX